LIAL
VIDEO WORKBOOK

CHRISTINE VERITY

INTRODUCTORY ALGEBRA
ELEVENTH EDITION

Margaret L. Lial
American River College

John Hornsby
University of New Orleans

Terry McGinnis

P Pearson

Copyright © 2018 Pearson Education, Inc.
Publishing as Pearson, 330 Hudson Street, NY NY 10013

ScoutAutomatedPrintCode

Pearson

ISBN-13: 978-0-13-450923-5
ISBN-10: 0-13-450923-4

CONTENTS

Chapter R PREALGEBRA REVIEW

R.1   Fractions

Learning Objectives
1   Identify prime numbers.
2   Write numbers in prime factored form.
3   Write fraction in lowest terms.
4   Convert between improper fractions and mixed numbers.
5   Multiply and divide fractions.
6   Add and subtract fractions.
7   Solve applied problems that involve fractions.
8   Interpret data from a circle graph.

Key Terms

Use the vocabulary terms listed below to complete each statement in exercises 1−9.

numerator	**denominator**	**proper fraction**
improper fraction	**equivalent fractions**	**lowest terms**
prime number	**composite number**	**prime factorization**

1.   Two fractions are _____ when they represent the same portion of a whole.

2.   A fraction whose numerator is larger than its denominator is called an _____.

3.   In the fraction $\frac{2}{9}$, the 2 is the _____.

4.   A fraction whose denominator is larger than its numerator is called a _____.

5.   The _____ of a fraction shows the number of equal parts in a whole.

6.   A _____ has at least one factor other than itself and 1.

7.   In a _____ every factor is a prime number.

8.   The factors of a _____ are itself and 1.

9.   A fraction is written in _____ when its numerator and denominator have no common factor other than 1.

Name: _____ Date: _____
Instructor: _____ Section: _____

Objective 1 Identify prime numbers.

Objective 1 Practice Exercises

For extra help, see Example 1 on page 2 of your text.

Tell whether each number is prime, composite, *or* neither.

1. 29

2. 35

3. 1

1. _____

2. _____

3. _____

Objective 2 Write numbers in prime factored form.

Video Examples

Review these examples for Objective 2:

2. Write each number in prime factored form.

   a. 26

   We factor using prime factors 2 and 13, as
   $26 = 2 \cdot 13$.

   b. 48

   We use a factor tree, as shown below. The prime
   factors are boxed.

   Divide by the
   least prime factor
   of 54, which is 2. $54 = 2 \cdot 27$

   Divide 27 by 3
   to find two $54 = 2 \cdot 3 \cdot 9$
   factors of 27.

   Now factor
   9 as $3 \cdot 3$. $54 = 2 \cdot 3 \cdot 3 \cdot 3$

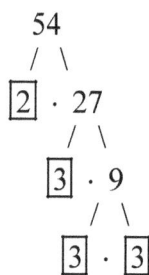

Now Try:

2. Write each number in prime factored form.

   a. 55

   _____

   b. 210

   _____

Objective 2 Practice Exercises

For extra help, see Example 2 on page 2 of your text.

Write each number in prime factored form.

4. 98

4. _____

5. 256   5. _____

6. 546   6. _____

Objective 3   Write fractions in lowest terms.

Video Examples

Review this example for Objective 3:
3c. Write the fraction in lowest terms.

$$\frac{75}{100}$$

$$\frac{75}{100} = \frac{3 \cdot 25}{4 \cdot 25} = \frac{3}{4} \cdot 1 = \frac{3}{4}$$

Now Try:
3c. Write the fraction in lowest terms.

$$\frac{81}{108}$$

_____

Objective 3 Practice Exercises

For extra help, see Example 3 on page 3 of your text.

Write each fraction in lowest terms.

7. $\dfrac{42}{150}$ 7. _____

8. $\dfrac{180}{216}$ 8. _____

9. $\dfrac{132}{292}$ 9. _____

Name: _____   Date: _____

Instructor: _____   Section: _____

Objective 4   Convert between improper fractions and mixed numbers.

Video Examples

Review these examples for Objective 4:

4. Write $\frac{53}{6}$ as a mixed number.

We divide the numerator of the improper fraction by the denominator.

$$\begin{array}{r} 8 \\ 6\overline{)53} \\ \underline{48} \\ 5 \end{array} \qquad \frac{53}{6} = 8\frac{5}{6}$$

5. Write $5\frac{3}{8}$ as an improper fraction.

We multiply the denominator of the fraction by the whole number and add the numerator to get the numerator of the improper fraction.

$$8 \cdot 5 + 3 = 40 + 3 = 43$$

The denominator of the improper fraction is the same as the denominator in the mixed number, which is 8 here. Thus, $5\frac{3}{8} = \frac{43}{8}$

Now Try:

4. Write $\frac{74}{5}$ as a mixed number.

_____

5. Write $12\frac{2}{7}$ as an improper fraction.

_____

Objective 4 Practice Exercises

For extra help, see Examples 4–5 on page 4 of your text.

Write the improper fraction as a mixed number.

10. $\frac{321}{15}$

10. _____

Write each mixed number as an improper fraction.

11. $13\frac{5}{9}$

11. _____

12. $22\frac{2}{11}$

12. _____

Name: Date:
Instructor: Section:

Objective 5   Multiply and divide fractions.

Video Examples

Review these examples for Objective 5:	**Now Try:**

Review these examples for Objective 5:

7. Find each quotient, and write it in lowest terms.

  b. $\dfrac{2}{5} \div \dfrac{8}{7}$

  $\dfrac{2}{5} \div \dfrac{8}{7} = \dfrac{2}{5} \cdot \dfrac{7}{8}$ Multiply by the reciprocal.

  $\quad = \dfrac{2 \cdot 7}{5 \cdot 4 \cdot 2}$ Multiply and factor.

  $\quad = \dfrac{7}{20}$

  c. $\dfrac{7}{9} \div 14$

  $\dfrac{7}{9} \div 14 = \dfrac{7}{9} \cdot \dfrac{1}{14}$ Multiply by the reciprocal.

  $\quad = \dfrac{7 \cdot 1}{9 \cdot 2 \cdot 7}$ Multiply and factor.

  $\quad = \dfrac{1}{18}$

Now Try:

7. Find each quotient, and write it in lowest terms.

  b. $\dfrac{6}{7} \div \dfrac{9}{8}$

  _____

  c. $\dfrac{4}{5} \div 8$

  _____

Objective 5 Practice Exercises

For extra help, see Examples 6–7 on pages 4–6 of your text.

Find each product or quotient, and write it in lowest terms.

13. $\dfrac{25}{11} \cdot \dfrac{33}{10}$

13. _____

14. $\dfrac{5}{4} \div \dfrac{25}{28}$

14. _____

15. $4\dfrac{3}{8} \cdot 2\dfrac{4}{7}$

15. _____

Objective 6   Add and subtract fractions.

Video Examples

Review these examples for Objective 6:

9a.   Add. Write sum in lowest terms.

$$\frac{5}{21} + \frac{3}{14}$$

Step 1 To find the LCD, factor the denominators to prime factored form.

    $21 = 3 \cdot 7$ and $14 = 2 \cdot 7$

7 is a factor of both denominators.

$$\begin{array}{cc} 21 & 14 \\ \wedge & \wedge \end{array}$$

Step 2 $LCD = 3 \cdot 7 \cdot 2 = 42$

In this example, the LCD needs one factor of 3, one factor of 7 and one factor of 2.

Step 3 Now we can use the second property of 1 to write each fraction with 42 as the denominator.

$$\frac{5}{21} = \frac{5}{21} \cdot \frac{2}{2} = \frac{10}{42} \quad \text{and} \quad \frac{3}{14} = \frac{3}{14} \cdot \frac{3}{3} = \frac{9}{42}$$

Now add the two equivalent fractions to get the sum.

$$\frac{5}{21} + \frac{3}{14} = \frac{10}{42} + \frac{9}{42}$$
$$= \frac{19}{42}$$

10b.   Subtract. Write difference in lowest terms.

$$\frac{14}{15} - \frac{5}{8}$$

Since 15 and 8 have no common factors greater than 1, the LCD is $15 \cdot 8 = 120$.

$$\frac{14}{15} - \frac{5}{8} = \frac{14}{15} \cdot \frac{8}{8} - \frac{5}{8} \cdot \frac{15}{15}$$
$$= \frac{112}{120} - \frac{75}{120}$$
$$= \frac{37}{120}$$

Now Try:

9a.   Add. Write sum in lowest terms.

$$\frac{7}{12} + \frac{3}{8}$$

_____

10b.   Subtract. Write difference in lowest terms.

$$\frac{17}{6} - \frac{13}{7}$$

_____

Objective 6 Practice Exercises

For extra help, see Example 8–10 on pages 7–10 of your text.

Find each sum or difference, and write it in lowest terms.

16. $\dfrac{23}{45}+\dfrac{47}{75}$ 16. _____

17. $2\dfrac{3}{4}+7\dfrac{2}{3}$ 17. _____

18. $12\dfrac{5}{6}-7\dfrac{7}{8}$ 18. _____

Objective 7   Solve applied problems that involve fractions.

Video Examples

Review this example for Objective 7:	**Now Try:**
12. A pie requires $3\dfrac{1}{3}$ cups of apples. How many pies can be made with $20\dfrac{1}{2}$ cups of apples? To better understand the problem, we replace the fractions with whole numbers. Suppose each pie requires 3 cups apples and we have a total of 21 cups of apples. Dividing 21 by 3 gives 7, the number of pies that can be made. To solve the	**12.** A pumpkin pie requires $\dfrac{3}{4}$ cup sugar. A large container of sugar has $10\dfrac{2}{3}$ cups of sugar. How many pumpkin pies can be made with this sugar? _____

original problem, we must divide $20\frac{1}{2}$ by $3\frac{1}{3}$.

Convert the mixed numbers to improper fractions. Then multiply by the reciprocal.

$$20\frac{1}{2} \div 3\frac{1}{3} = \frac{41}{2} \div \frac{10}{3}$$
$$= \frac{41}{2} \cdot \frac{3}{10}$$
$$= \frac{123}{20}, \text{ or } 6\frac{3}{20}$$

Thus, 6 pies can be made with some apples left over.

Objective 7 Practice Exercises

For extra help, see Examples 11–12 on pages 10–11 of your text.

Solve each applied problem. Write each answer in lowest terms.

19.   Arnette worked $24\frac{1}{2}$ hours and earned \$9 per hour. How much did she earn?   **19.** _____

20.   Debbie bought 15 yards of material at a sale. She made a shirt with $3\frac{1}{8}$ yards of the material, a dress with $4\frac{7}{8}$ yards, and a jacket with $3\frac{3}{4}$ yards. How many yards of material were left over?   **20.** _____

21. Three sides of a parking lot are $35\frac{1}{4}$ yards, $42\frac{7}{8}$ **21.** _____

yards, and $32\frac{3}{4}$ yards. If the total distance around

the lot is $145\frac{1}{2}$ yards, find the length of the fourth

side.

Objective 8   Interpret data from a circle graph.

Video Examples

In August 2016, 1300 workers were surveyed on where they eat during their lunch time. The circle graph shows the approximate fractions of locations.

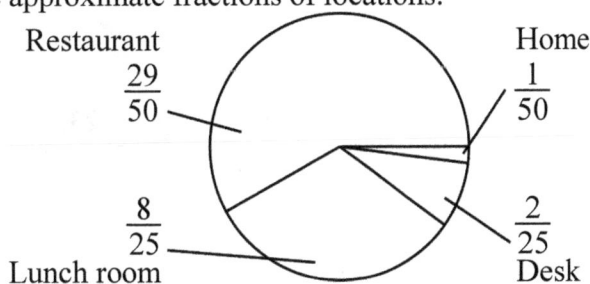

Restaurant $\frac{29}{50}$ Home $\frac{1}{50}$

$\frac{8}{25}$ Lunch room $\frac{2}{25}$ Desk

Review this example for Objective 8:	**Now Try:**
13c.	**13c.**

How many actual workers ate in the Lunch Room?

Multiply the actual fraction from the graph of Lunch Room by the number of workers surveyed.

$$\frac{8}{25} \cdot 1300 = \frac{8}{25} \cdot \frac{1300}{1}$$
$$= \frac{10,400}{25}$$
$$= 416$$

Thus, 416 workers ate in the Lunch Room.

How many actual workers ate at their Desk?

_____

Name: _____ Date: _____

Instructor: _____ Section: _____

Objective 8 Practice Exercises

For extra help, see Example 13 on page 11 of your text.

In August 2016, 1300 workers were surveyed on where they eat during their lunch time. The circle graph shows the approximate fractions of locations.

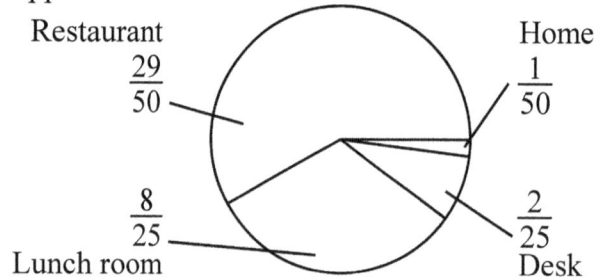

Restaurant $\frac{29}{50}$

Home $\frac{1}{50}$

$\frac{8}{25}$ Lunch room

$\frac{2}{25}$ Desk

22. What location had the second-largest number of workers?

22. _____

23. Estimate the number of workers who eat at a Restaurant.

23. _____

24. How many actual workers ate at a Restaurant?

24. _____

Chapter R PREALGEBRA REVIEW

R.2   Decimals and Percents

Learning Objectives
1   Write decimals as fractions.
2   Add and subtract decimals.
3   Multiply and divide decimals.
4   Write fractions as decimals.
5   Write percents as decimals and decimals as percents.
6   Write percents as fractions and fractions as percents.
7   Solve applied problems that involve percents.

Key Terms

Use the vocabulary terms listed below to complete each statement in exercises 1−3.

decimals   place value   percent

1.   We use _____ to show parts of a whole.

2.   A _____ is assigned to each place to the left or right of the decimal point.

3.   _____ means per one hundred.

Objective 1   Write decimals as fractions.

Video Examples

Review this example for Objective 1:

1c. Write the decimal as a fraction. Do not write in lowest terms.

5.3084

Here we have 4 places.
$$5.3084 = 5 + 0.3084$$

$$= \frac{50,000}{10,000} + \frac{3084}{10,000} \quad \text{The LCD is 10,000.}$$

$$= \frac{53,084}{10,000}$$

4 zeros

Now Try:

1c. Write the decimal as a fraction. Do not write in lowest terms.

3.7058

_____

Objective 1 Practice Exercises

For extra help, see Example 1 on pages 17–18 of your text.

Write each decimal as a fraction. Do not write in lowest terms.

   1.   0.007   **1.** _____

   2.   18.03   **2.** _____

   3.   30.0005   **3.** _____

Objective 2   Add and subtract decimals.

Video Examples

Review this example for Objective 2:

2c. Add or subtract as indicated.

    $5 - 0.832$

    A whole number is assumed to have the decimal point at the right of the number. Write 5 as 5.000.

$$\begin{array}{r} 5.000 \\ -\ 0.832 \\ \hline 4.168 \end{array}$$

Now Try:

2c. Add or subtract as indicated.

    $8 - 0.976$

    _____

Objective 2 Practice Exercises

For extra help, see Example 2 on page 18 of your text.

Add or subtract as indicated.

   4. $45.83 + 20.923 + 5.7$ **4.** _____

   5. $768.5 - 13.402$ **5.** _____

   6. $689 - 79.832$ **6.** _____

Objective 3   Multiply and divide decimals.

Video Examples

Review these examples for Objective 3:

3a. Multiply.

$$37.4 \times 5.26$$

There is 1 decimal place in the first number and 2 decimal places in the second number. Therefore, there are $1 + 2 = 3$ decimal places in the answer.

$$
\begin{array}{r}
37.4 \\
\times\ 5.26 \\
\hline
2244 \\
748 \\
1870 \\
\hline
196.724
\end{array}
$$

4a. Divide.

$$1191.45 \div 23.5$$

Write the problem as follows.

$$23.5\overline{)1191.45}$$

To change 23.5 into a whole number, move the decimal point one place to the right. Move the decimal point in 1191.45 the same number of places to the right, to get 11,914.5.

$$235\overline{)11,914.5}$$

Move the decimal point straight up and divide as with whole numbers.

$$
\begin{array}{r}
50.7 \\
235\overline{)11,914.5} \\
\underline{1175} \\
1645 \\
\underline{1645} \\
0
\end{array}
$$

5. Multiply or divide as indicated.

a. 97.648×100

Move the decimal point 2 places to the right because 100 has two zeros.
$$97.648 \times 100 = 9764.8$$

Now Try:

3a. Multiply.

$$26.8 \times 9.37$$

_____

4a. Divide.

$$2472.12 \div 76.3$$

_____

5. Multiply or divide as indicated.

a. 51.302×100

_____

b. $85.1 \div 1000$

Move the decimal point three places to the left because 1000 has three zeros. Insert a zero in front of the 8 to do this.

$85.1 \div 1000 = 0.0851$

b. $98.6 \div 1000$

_____

Objective 3 Practice Exercises

For extra help, see Examples 3–5 on pages 19–20 of your text.

Multiply or divide as indicated.

7. 14.64×0.16 7. _____

8. $498.624 \div 21.2$ 8. _____

9. $429.2 \div 1000$ 9. _____

Objective 4   Write fractions as decimals.

Video Examples

Review these examples for Objective 4:

6. Write each fraction as a decimal. For repeating decimals, write the answer by first using bar notation and then rounding to the nearest thousandth.

a. $\dfrac{27}{8}$

Divide 27 by 8. Add a decimal point and as many 0s as necessary.

$$\begin{array}{r} 3.375 \\ 8\overline{)27.000} \\ \underline{24} \\ 30 \\ \underline{24} \\ 60 \\ \underline{56} \\ 40 \\ \underline{40} \\ 0 \end{array}$$

$\dfrac{27}{8} = 3.375$

b. $\dfrac{26}{9}$

$$\begin{array}{r} 2.888... \\ 9\overline{)26.000...} \\ \underline{18} \\ 80 \\ \underline{72} \\ 80 \\ \underline{72} \\ 80 \\ \underline{72} \\ 8 \end{array}$$

$\dfrac{26}{9} = 2.888...$

The remainder is never 0. Because 8 is always left after the subtraction, this quotient is a repeating decimal. A convenient notation for a repeating decimal is a bar over the digit (or digits) that repeats.

$\dfrac{26}{9} = 2.\overline{8}$ or $\dfrac{26}{9} \approx 2.889$ rounded

Now Try:

6. Write each fraction as a decimal. For repeating decimals, write the answer by first using bar notation and then rounding to the nearest thousandth.

a. $\dfrac{7}{20}$

_____

b. $\dfrac{32}{9}$

_____

Objective 4 Practice Exercises

For extra help, see Example 6 on page 20 of your text and Section Lecture video for Section R.2 and Exercise Solutions Clip 39 and 43.

Write each fraction as a decimal. For repeating decimals, write the answer two ways: using the bar notation and rounding to the nearest thousandth.

10. $\dfrac{3}{7}$ 10. _____

11. $\dfrac{4}{9}$ 11. _____

12. $\dfrac{151}{200}$ 12. _____

Objective 5   Write percents as decimals and decimals as percents.

Video Examples

Review these examples for Objective 5:

9. Convert each percent to a decimal and each decimal to a percent.

   c. 0.29%

   0.29% = 0.0029

   d. 0.54

   0.54 = 54%

Now Try:

9. Convert each percent to a decimal and each decimal to a percent.

   c. 0.43%

                _____

   d. 0.91

                _____

Objective 5 Practice Exercises

For extra help, see Examples 7–9 on pages 21–22 of your text.

Convert each percent to a decimal and each decimal to a percent.

13.   362%   **13.** _____

14.   0.4%   **14.** _____

15.   0.0084   **15.** _____

Objective 6   **Write percents as fractions and fractions as percents.**

Video Examples

Review this example for Objective 6:

10a. Write each percent as a fraction. Give answers in lowest terms as needed.

65%

We use the fact that, $1\% = \dfrac{1}{100}$, and convert as follows.

$$65\% = 65 \cdot 1\% = 65 \cdot \frac{1}{100} = \frac{65}{100}$$

In lowest terms,

$$\frac{65}{100} = \frac{13 \cdot 5}{20 \cdot 5} = \frac{13}{20}.$$

Thus,

$$65\% = \frac{13}{20}.$$

Now Try:

10a. Write each percent as a fraction. Give answers in lowest terms as needed.

52%

_____

Objective 6 Practice Exercises

For extra help, see Examples 10–11 on page 23 of your text.

Write each percent as a fraction. Give answers in lowest terms as needed.

16.   42%   **16.** _____

17.   3.5%   **17.** _____

18.   160%   18. _____

Objective 7   Solve applied problems that involve percents.

Video Examples

Review this example for Objective 6:

12. Ranee bought a pair of shows with a regular price of $70, on sale at 20% off. How much money did she save? What is the sale price?

From the table in the text, $20\% = \frac{1}{5}$ and $\frac{1}{5}$ of $70 is $14, so she will save $14.

Original price – discount = sale price
    $70 – $14 = $56

Now Try:

12. At the end of the season, a swim suit is on sale at 75% off. The regular price is $56. What is the sale price? How much is saved?

_____

Objective 6 Practice Exercises

For extra help, see Example 12 on page 24 of your text.

Solve each problem.

19.   A television set sells for $750 plus 8% sales tax. Find the price of the set including sales tax.   19. _____

20.   Geishe's Shoes sells shoes at $33\frac{1}{3}\%$ off the regular price. Find the price of a pair of shoes normally priced at $54, after the discount is given.   20. _____

21.   A house costs $225,000. The Lee's paid $45,000 as a down payment. What percent of the cost of the house is their down payment?   21. _____

Chapter 1 THE REAL NUMBER SYSTEM

1.1   Exponents, Order of Operations, and Inequality

Learning Objectives
1   Use exponents.
2   Use the rules for order of operations.
3   Use more than one grouping symbol.
4   Know the meanings of $\neq$, $<$, $>$, $\leq$, and $\geq$.
5   Translate word statements to symbols.
6   Write statements that change the direction of inequality symbols.

Key Terms

Use the vocabulary terms listed below to complete each statement in exercises 1–4.

exponent   base   exponential expression   inequality

1.   A number written with an exponent is an _____.

2.   The _____ is the number that is a repeated factor when written with an exponent.

3.   An _____ is a statement that two expressions may not be equal.

4.   An _____ is a number that indicates how many times a factor is repeated.

Objective 1   Use exponents.

Video Examples

Review this example for Objective 1:
1a. Find the value of the exponential expression.

$$6^2$$

6^2 means $6 \cdot 6$, which equals 36.

Now Try:
1a. Find the value of the exponential expression.
9^2

_____

Objective 1 Practice Exercises

For extra help, see Example 1 on page 30 of your text.

Find the value of each exponential expression.

1. 3^3

2. $\left(\dfrac{2}{3}\right)^4$

3. $(0.4)^2$

1.   _____

2.   _____

3.   _____

Objective 2   Use the rules for order of operations.

Video Examples

Review this example for Objective 2:
2a. Find the value of the expression.

$$56 - 32 \div 8$$

Divide, then subtract.
$$56 - 32 \div 8 = 56 - 4$$
$$= 52$$

Now Try:
2a. Find the value of the expression.

$$93 - 56 \div 7$$

_____

Objective 2 Practice Exercises

For extra help, see Example 2 on pages 31–32 of your text.

Find the value of each expression.

4. $20 \div 5 - 3 \cdot 1$

4. _____

5. $3 \cdot 5^2 - 3 \cdot 7 - 9$

5. _____

6. $6^2 \div 3^2 - 4 \cdot 3 - 2 \cdot 5$

6. _____

Objective 3   Use more than one grouping symbol.

Video Examples

Review this example for Objective 3:
3a. Find the value of the expression.

$$3[9 + 4(7 + 8)]$$

Start by adding inside the parentheses.
$$3[9 + 4(7 + 8)] = 3[9 + 4(15)] \quad \text{Add.}$$
$$= 3[9 + 60] \quad \text{Multiply.}$$
$$= 3[69] \quad \text{Add.}$$
$$= 207 \quad \text{Multiply.}$$

Now Try:
3a. Find the value of the expression.

$$5[3 + 4(8 + 2)]$$

_____

Objective 3 Practice Exercises

For extra help, see Example 3 on page 32 of your text.

Find the value of each expression.

7. $\dfrac{10(5-3)-9(6-2)}{2(4-1)-2^2}$

7. _____

8. $19-3\big[8(5-2)+6\big]$

8. _____

9. $4\big[5+2(8-6)\big]+12$

9. _____

Objective 4   Know the meanings of $\neq, <, >, \leq,$ and $\geq$.

Video Examples

Review this example for Objective 4:
4c. Determine whether the statement is true or false.

$14 \leq 50 \cdot 3$

The statement $14 \leq 50 \cdot 3$ is true, because $14 < 150$.

Now Try:
4c. Determine whether the statement is true or false.
$25 \leq 49 \cdot 2$

_____

Objective 4 Practice Exercises

For extra help, see Example 4 on page 33 of your text.

Tell whether each statement is true *or* false.

10. $3 \cdot 4 \div 2^2 \neq 3$

10. _____

11. $3.25 > 3.52$ **11.** _____

12. $2\left[7(4) - 3(5)\right] \le 45$ **12.** _____

Objective 5   Translate word statements to symbols.

Video Examples

Review this example for Objective 5:	**Now Try:**
5d. Write each word statement in symbols.	**5d.** Write each word statement in symbols.
Eight is greater than six.	Five is greater than three.
$8 > 6$	_____

Objective 5 Practice Exercises

For cxtra help, see Example 5 on page 34 of your text.

Write each word statement in symbols.

13.   Seven equals thirteen minus six.   **13.** _____

14.   Five times the sum of two and nine is less than one hundred six.   **14.** _____

15.   Twenty is greater than or equal to the product of two and seven.   **15.** _____

Objective 6   Write statements that change the direction of inequality symbols.

Video Examples

Review this example for Objective 6:

6a. Write each statement as another true statement with the inequality symbol reversed.

$9 > 7$

$7 < 9$

Now Try:

6a. Write each statement as another true statement with the inequality symbol reversed.
$15 > 11$

_____

Objective 6 Practice Exercises

For extra help, see Example 6 on page 34 of your text.

Write each statement with the inequality symbol reversed.

16. $\dfrac{3}{4} > \dfrac{2}{3}$

16. _____

17. $12 \geq 8$

17. _____

18. $0.002 > 0.0002$

18. _____

Chapter 1 THE REAL NUMBER SYSTEM

1.2   Variables, Expressions, and Equations

Learning Objectives
1   Evaluate algebraic expressions, given values for the variables.
2   Translate word phrases to algebraic expressions.
3   Identify solutions of equations.
4   Translate sentences to equations.
5   Distinguish between equations and expressions.

Key Terms

Use the vocabulary terms listed below to complete each statement in exercises 1−5.

variable	constant	algebraic expression
equation	solution	

1.   A(n) _____ is a statement that says two expressions are equal.

2.   A _____ is a symbol, usually a letter, used to represent an unknown number.

3.   A collection of numbers, variables, operation symbols, and grouping symbols is an_____.

4.   Any value of a variable that makes an equation true is a(n) _____ of the equation.

5.   A _____ is a fixed, unchanging number.

Objective 1   Evaluate algebraic expressions, given values for the variables.

Video Examples

Review these examples for Objective 1:	Now Try:
1d. Find the value of each algebraic expression for $p = 4$ and then $p = 7$.	**1d.** Find the value of each algebraic expression for $k = 6$ and then $k = 9$.
$5p^2$	
For $p = 4$,	$7k^2$
$\quad 5p^2 = 5 \cdot 4^2$ Let $p = 4$.	
$\quad\quad = 5 \cdot 16$ Square 4.	
$\quad\quad = 80$ Multiply.	_____
For $p = 7$,	
$\quad 5p^2 = 5 \cdot 7^2$ Let $p = 7$.	
$\quad\quad = 5 \cdot 49$ Square 7.	
$\quad\quad = 245$ Multiply.	

Name: Date:

Instructor: Section:

2. Find the value of each expression for $x = 7$ and $y = 6$.

    a. $3x + 4y$

    Replace x with 7 and y with 6.

$$3x + 4y = 3 \cdot 7 + 4 \cdot 6$$
$$= 21 + 24 \qquad \text{Multiply.}$$
$$= 45 \qquad \text{Add.}$$

    b. $\dfrac{8x - 6y}{4x - 3y}$

    Replace x with 7 and y with 6.

$$\frac{8x - 6y}{4x - 3y} = \frac{8 \cdot 7 - 6 \cdot 6}{4 \cdot 7 - 3 \cdot 6}$$
$$= \frac{56 - 36}{28 - 18} \qquad \text{Multiply.}$$
$$= \frac{20}{10} \qquad \text{Subtract.}$$
$$= 2 \qquad \text{Divide.}$$

2. Find the value of each expression for $x = 8$ and $y = 4$.

    a. $5x + 6y$

    _____

    b. $\dfrac{9x + 2y}{3x - 5y}$

    _____

Objective 1 Practice Exercises

For extra help, see Examples 1–2 on pages 39–40 of your text.

Find the value of each expression if $x = 2$ and $y = 4$.

1. $9x - 3y + 2$

1. _____

2. $\dfrac{2x + 3y}{3x - y + 2}$

2. _____

3. $\dfrac{3y^2 + 2x^2}{5x + y^2}$

3. _____

Objective 2   Translate word phrases to algebraic expressions.

Video Examples

Review these examples for Objective 2:

3. Write each word phrase as an algebraic expression, using x as the variable.

    c. A number subtracted from 13

    $13 - x$

    f. The product of 3 and the difference between a number and 5

    $3 \cdot (x - 5)$, or $3(x - 5)$

Now Try:

3. Write each word phrase as an algebraic expression, using x as the variable.

    c. A number subtracted from 12

    _____

    f. The product of 5 and the difference between a number and 6

    _____

Objective 2 Practice Exercises

For extra help, see Example 3 on pages 40–41 of your text.

Write each word phrase as an algebraic expression. Use x as the variable.

4.   Ten times a number, added to 21   4. _____

5.   11 fewer than eight times a number   5. _____

6.   Half a number subtracted from two-thirds of the number   6. _____

Objective 3   Identify solutions of equations.

Video Examples

Review these examples for Objective 3:

4. Decide whether the given number is a solution of the equation.

    a. $6p + 5 = 29;\quad 4$

    $6p + 5 = 29$
    $6 \cdot 4 + 5 \overset{?}{=} 29$
    $24 + 5 \overset{?}{=} 29$
    $29 = 29$ True – the left side of the equation equals the right side.

    The number 4 is a solution of the equation.

Now Try:

4. Decide whether the given number is a solution of the equation.

    a. $7k + 5 = 26;\quad 3$

    _____

b. $8n - 7 = 41;$ $\dfrac{9}{4}$ **b.** $9m - 8 = 41;$ 5

$8n - 7 = 41$

$8 \cdot \overset{?}{\dfrac{9}{4}} - 7 = 41$ _____

$18 \overset{?}{-} 7 = 41$

$\quad 11 = 41$ False – the left side does not
$\qquad\qquad$ equal the right side.

The number $\dfrac{9}{4}$ is not a solution of the equation.

Objective 3 Practice Exercises

For extra help, see Example 4 on page 41 of your text.

Decide whether the given number is a solution of the equation.

7. $5 + 3x^2 = 19;$ 2   7. _____

8. $\dfrac{m + 2}{3m - 10} = 1;$ 8   8. _____

9. $3y + 5(y - 5) = 7;$ 4   9. _____

Objective 4   Translate sentences to equations.

Video Examples

Review this example for Objective 4:   **Now Try:**

5c. Write the word sentence as an equation.   **5c.** Write the word sentence as an
Use x as the variable.   equation. Use x as the variable.

Nine less than five times a number is equal to   Seven less than three times a
twenty.   number is equal to twelve.

Five times
a number   less   nine   is equal to   twenty.   _____
$\quad\downarrow$ $\downarrow$ $\downarrow$ $\downarrow$ $\downarrow$
$\quad 5x$ $-$ 9 $=$ 20

$\qquad\qquad 5x - 9 = 20$

Objective 4 Practice Exercises

For extra help, see Example 5 on page 42 of your text.

Write each word sentence as an equation. Use x as the variable.

10. Ten divided by a number is two more than the number.

10. _____

11. The product of six and five more than a number is nineteen.

11. _____

12. Seven times a number subtracted from 61 is 13 plus the number.

12. _____

Objective 5 Distinguish between equations and expressions.

Video Examples

Review these examples for Objective 5:

6. Decide whether each is an equation or an expression.

   a. $5x - 4$

   Ask, "Is there an equality symbol?" The answer is no, so this is an expression.

   b. $12x - 11 = 7$

   Because there is an equality symbol with something on either side of it, this is an equation.

Now Try:

6. Decide whether each is an equation or an expression.

   a. $4y - 6 = 20$

   _____

   b. $\dfrac{3x - 15y}{2}$

   _____

Objective 5 Practice Exercises

For extra help, see Example 6 on page 42 of your text.

*Identify each as an **expression** or an **equation**.*

13. $y^2 - 4y - 3$

13. _____

14. $\dfrac{x + 4}{5}$

14. _____

15. $8x = 2y$

15. _____

Name: Date:
Instructor: Section:

Chapter 1 THE REAL NUMBER SYSTEM

1.3   Real Numbers and the Number Line

Learning Objectives	
1	Classify numbers and graph them on number lines.
2	Tell which of two real numbers is less than the other.
3	Find the additive inverse of a real number.
4	Find the absolute value of a real number.

Key Terms

Use the vocabulary terms listed below to complete each statement in exercises 1–14.

natural numbers	**whole numbers**	**number line**	**additive inverse**
integers	**negative number**	**positive number**	**signed numbers**
rational number	**set-builder notation**		**coordinate**
irrational number	**real numbers**	**absolute value**	

1.   The set {0, 1, 2, 3, …} is called the set of _____.

2.   The _____ of a number is the same distance from 0 on the number line as the original number, but located on the opposite side of 0.

3.   The whole numbers together with their opposites and 0 are called _____.

4.   The set { 1, 2, 3, …} is called the set of _____.

5.   The _____ of a number is the distance between 0 and the number on the number line.

6.   A _____ shows the ordering of the real numbers on a line.

7.   A real number that is not a rational number is called a(n) _____.

8.   The number that corresponds to a point on the number line is the _____ of that point.

9.   A number located to the left of 0 on a number line is a _____.

10.   A number located to the right of 0 on a number line is a _____.

11.   Numbers that can be represented by points on the number line are _____.

12.   _____ uses a variable and a description to describe a set.

13.   Positive numbers and negative numbers are _____.

14. A number that can be written as the quotient of two integers is a

_____.

Objective 1   Classify numbers and graph them on number lines.

Video Examples

Review these examples for Objective 1:	Now Try:

Review these examples for Objective 1:

1a. Use an integer to express the boldface italic number in the application.

  a. In August, 2012, the National Debt was approximately $*16* trillion.

  Use –$16 trillion because "debt" indicates a negative number.

2. Graph each number on a number line.

$-3\frac{1}{2}$, $-\frac{3}{2}$, 0, $\frac{7}{2}$, 1

To locate the improper fractions on the number line, write them as mixed numbers or decimals.

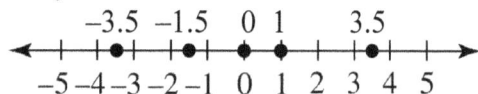
–3.5 –1.5   0 1   3.5
–5 –4 –3 –2 –1 0 1 2 3 4 5

3. List the numbers in the following set that belong to each set of numbers.

$\left\{-6, -\frac{5}{6}, 0, 0.\overline{3}, \sqrt{3}, 4\frac{1}{5}, 6, 6.7\right\}$

  a. Natural numbers

  Answer: 6

  b. Whole numbers

  Answer: 0 and 6

  c. Integers

  Answer: –6, 0, and 6

  d. Rational numbers

  Answer: $-6, -\frac{5}{6}, 0, 0.\overline{3}, 4\frac{1}{5}, 6, 6.7$

  e. Irrational numbers

  Answer: $\sqrt{3}$

  f. Real numbers

  Answer: all the numbers in the set

Now Try:

1a. Use an integer to express the boldface italic number in the application.
  a. Death Valley is *282* feet below sea level.

  _____

2. Graph each number on a number line.

$\frac{1}{2}$, 0, -3, $-\frac{5}{2}$

–5 –4 –3 –2 –1 0 1 2 3 4 5

3. List the numbers in the following set that belong to each set of numbers.

$\left\{-10, -\frac{5}{8}, 0, 0.\overline{4}, \sqrt{5}, 5\frac{1}{2}, 7, 9.9\right\}$

  a. Natural numbers

  _____

  b. Whole numbers

  _____

  c. Integers

  _____

  d. Rational numbers

  _____

  e. Irrational numbers

  _____

  f. Real numbers

  _____

Name: Date:
Instructor: Section:

Objective 1 Practice Exercises

For extra help, see Examples 1–3 on pages 48–50 of your text.

Use a real number to express each number in the following applications.

1. Last year Nina lost 75 pounds.   1. _____

2. Between 1970 and 1982, the population of Norway   2. _____
    increased by 279,867.

Graph the group of rational numbers on a number line.

3. −4.5, −2.3, 1.7, 4.2   3.

−5 −4 −3 −2 −1 0 1 2 3 4 5

Objective 2   Tell which of two real numbers is less than the other.

Video Examples

Review this example for Objective 2:

4. Is the statement $-4 < -2$ true or false?

    Because −4 is to the left of −2 on the number line, −4 is less than −2. The statement $-4 < -2$ is true.

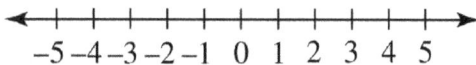

−5 −4 −3 −2 −1 0 1 2 3 4 5

Now Try:

4. Is the statement $-10 < -8$ true or false.

    _____

Objective 2 Practice Exercises

For extra help, see Example 4 on page 51 of your text.

*Decide whether each statement is **true** or **false**.*

4. $-76 < 45$ 4. _____

5. $-5 > -5$ 5. _____

6. $-12 > -10$ 6. _____

Objective 3   Find the additive inverse of a real number.

Objective 3 Practice Exercises

For extra help, see page 51 of your text.

Find the additive inverse of each number.

7. -25 7. _____

8. $\dfrac{3}{8}$ 8. _____

9. 4.5 9. _____

Objective 4   Find the absolute value of a real number.

Video Examples

Review these examples for Objective 4:	**Now Try:**
5. Simplify by finding the absolute value.	5. Simplify by finding the absolute value.
b. $\lvert 16 \rvert$	**b.** $\lvert 10 \rvert$
$\lvert 16 \rvert = 16$	_____
c. $\lvert -16 \rvert$	**c.** $\lvert -10 \rvert$
$\lvert -16 \rvert = -(-16) = 16$	_____
e. $-\lvert -16 \rvert$	**e.** $-\lvert -10 \rvert$
$-\lvert -16 \rvert = -(16) = -16$	_____

Objective 4 Practice Exercises

For extra help, see Example 5 on page 52 of your text.

Simplify.

10. $-\lvert 49 - 39 \rvert$ 10. _____

11. $\lvert -7.52 + 6.3 \rvert$ 11. _____

12. $\lvert 16 - 14 \rvert$ 12. _____

Chapter 1 THE REAL NUMBER SYSTEM

1.4   Adding Real Numbers

Learning Objectives
1   Add two numbers with the same sign.
2   Add numbers with different signs.
3   Use the rules for order of operations when adding real numbers.
4   Translate words and phrases that indicate addition.

Key Terms

Use the vocabulary terms listed below to complete each statement in exercises 1−2.

         sum   **addends**

1.   The answer to an addition problem is called the _____.

2.   In an addition problem, the numbers being added are the _____.

Objective 1   Add two numbers with the same sign.

Video Examples

Review these examples for Objective 1:

1.   Use a number line to find each sum.

     a. $4 + 4$

Step 1 Start at 0 and draw an arrow 4 units to the right.

Step 2 From the right end of that arrow, draw another arrow 4 units to the right.

The number below the end of this second arrow is 8, so $4 + 4 = 8$.

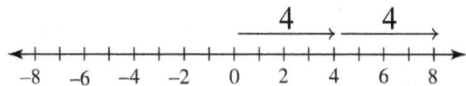

     b. $-3 + (-5)$

Step 1 Start at 0 and draw an arrow 3 units to the left.

Step 2 From the left end of that arrow, draw another arrow 5 units to the left.

The number below the end of this second arrow is −5, so $-3 + (-5) = -8$.

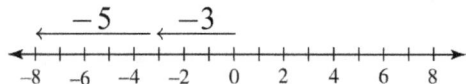

Now Try:

1.   Use a number line to find each sum.

     a. $2 + 4.$

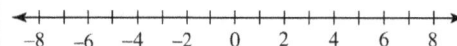

     _____

     b. $-4 + (-1)$

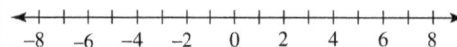

     _____

2. Find each sum.

   a. $-3 + (-7)$

   $-3 + (-7) = -10$

   b. $-5 + (-16)$

   $-5 + (-16) = -21$

   c. $-18 + (-7)$

   $-18 + (-7) = -25$

2. Find each sum.

   a. $-8 + (-4)$

   _____

   b. $-17 + (-14)$

   _____

   c. $-10 + (-30)$

   _____

Objective 1 Practice Exercises

For extra help, see Examples 1–2 on pages 57–58 of your text.

Find each sum.

  1. $-7 + (-11)$

1. _____

  2. $-9 + (-9)$

2. _____

  3. $-2\frac{3}{8} + \left(-3\frac{1}{4}\right)$

3. _____

Objective 2   Add numbers with different signs.

Video Examples

Review these examples for Objective 2:

4. Find each sum.

   a. $-10 + 6$

   Find the absolute value of each number.

     $|-10| = 10$ and $|6| = 6$

   Then find the difference between these absolute values: $10 - 6 = 4$. The sum will be negative since $|-10| > |6|$.

     $-10 + 6 = -4$

Now Try:

4. Find each sum.

   a. $-25 + 13$

   _____

b. $-16+9$

Find the absolute value of each number.
$$|-16|=16 \quad \text{and} \quad |9|=9$$
Then find the difference between these absolute values: $16-9=7$. The sum will be positive since $|-16|>|9|$.
$$-16+9=-7$$

b. $-12+5$

_____

Objective 2 Practice Exercises

For extra help, see Examples 3–4 on pages 58–59 of your text.

Use a number line to find the sum.

4. $-8+5$

4. _____

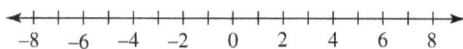

Find each sum.

5. $\frac{7}{12}+\left(-\frac{3}{4}\right)$

5. _____

6. $-7.5+9.4$

6. _____

Objective 3   Use the rules for order of operations when adding real numbers.

Video Examples

Review this example for Objective 3:
5b. Find the sum.

b. $9+\left[(-3+7)+(-5)\right]$

$$9+\left[(-3+7)+(-5)\right]=9+\left[4+(-5)\right]$$
$$=9+(-1)$$
$$=8$$

Now Try:
5b. Find the sum.

b. $19+\left[(-5+8)+(-7)\right]$

_____

Objective 3 Practice Exercises

For extra help, see Example 5 on pages 59–60 of your text.

Find each sum.

7. $-2+\left[4+(-18+13)\right]$ 7. _____

8. $\left[(-7)+14\right]+\left[(-16)+3\right]$ 8. _____

9. $-8.9+\left[6.8+(-4.7)\right]$ 9. _____

Objective 4   Translate words and phrases that indicate addition.

Video Examples

Review these examples for Objective 4:
6.   Write a numerical expression for each phrase, and simplify the expression.

   a. The sum of –9 and 5 and 3

   –9 + 5 + 3 simplifies to –4 + 3, which equals –1.

   b. 4 more than –7, increased by 15

   (–7 + 4) + 15 simplifies to –3 + 15, which equals 12.

Now Try:
6.   Write a numerical expression for each phrase, and simplify the expression.

   a. The sum of –10 and 11 and 2

   _____

   b. 15 more than –9, increased by 6

   _____

Objective 4 Practice Exercises

For extra help, see Examples 6–7 on page 60 of your text.

Write a numerical expression for each phrase, and then simplify the expression.

10. The sum of –14 and –29, increased by 27

10. _____

_____

11. –10 added to the sum of 20 and –4

11. _____

_____

Solve the problem.

12. The temperature at dawn in Blackwood was 24°F. During the day the temperature decreased 30°. Then it increased 11° by sunset. What was the temperature at sunset?

12. _____

Chapter 1 THE REAL NUMBER SYSTEM

1.5   Subtracting Real Numbers

Learning Objectives
1   Subtract two numbers on a number line.
2   Use the definition of subtraction.
3   Use the rules for order of operations when subtracting real numbers.
4   Translate words and phrases that indicate subtraction.

Key Terms

Use the vocabulary terms listed below to complete each statement in exercises 1−3.

minuend   **subtrahend**   **difference**

1.   The number from which another number is being subtracted is called the

_____.

2.   The _____ is the number being subtracted.

3.   The answer to a subtraction problem is called the _____.

Objective 1   Subtract two numbers on a number line.

Video Examples

Review this example for Objective 1:

1.   Use a number line to find the difference $5 - 3$.

Step 1 Start at 0 and draw an arrow 5 units to the right.

Step 2 From the right end of that arrow, draw another arrow 3 units to the left.

The number below the end of this second arrow is 2, so $5 - 3 = 2$.

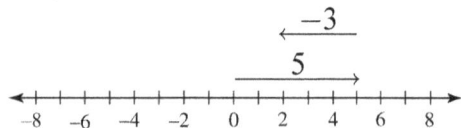

Now Try:

1.   Use a number line to find the difference $3 - 1$.

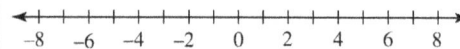

_____

Objective 1 Practice Exercises

For extra help, see Example 1 on page 64 of your text.

Use a number line to find the difference.

1. $8 - 5$

1. _____

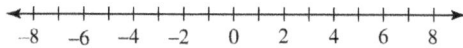

2. $7 - 10$

2. _____

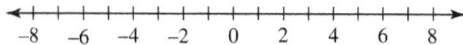

3. $-5 - 2$

3. _____

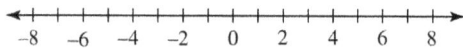

Objective 2   Use the definition of subtraction.

Video Examples

Review these examples for Objective 2:

2. Find each difference.

   b. $8 - 11$

   $8 - 11 = 8 + (-11) = -3$

   c. $-9 - 16$

   $-9 - 16 = -9 + (-16) = -25$

   d. $-6 - (-9)$

   $-6 - (-9) = -6 + (9) = 3$

   e. $\dfrac{5}{6} - \left(-\dfrac{3}{7}\right)$

   $\dfrac{5}{6} - \left(-\dfrac{3}{7}\right) = \dfrac{35}{42} - \left(-\dfrac{18}{42}\right)$

   $= \dfrac{35}{42} + \dfrac{18}{42}$

   $= \dfrac{53}{42}$

Now Try:

2. Find each difference.

   b. $13 - 17$

   _____

   c. $-11 - 27$

   _____

   d. $-5 - (-7)$

   _____

   e. $\dfrac{5}{9} - \left(-\dfrac{4}{5}\right)$

   _____

Objective 2 Practice Exercises

For extra help, see Example 2 on page 65 of your text.

Find each difference.

   4. $22-(-24)$ **4.** _____

   5. $-3.2-(-7.6)$ **5.** _____

   6. $\dfrac{15}{4}-\left(-\dfrac{17}{8}\right)$ **6.** _____

Objective 3   Use the rules for order of operations when subtracting real numbers.

Video Examples

Review these examples for Objective 3:	**Now Try:**
3. Perform each operation.	**3.** Perform each operation.

a. $-7-[3-(9+6)]$ **a.** $-10-[5-(7+3)]$

$$-7-[3-(9+6)]=-7-[3-15]$$
$$=-7-[3+(-15)]$$
$$=-7-(-12)$$
$$=-7+12$$
$$=5$$

b. $4-\left[\left(-\dfrac{1}{5}-\dfrac{1}{4}\right)-(3-2)\right]$ **b.** $8-\left[\left(-\dfrac{1}{6}-\dfrac{1}{3}\right)-(5-4)\right]$

$$4-\left[\left(-\dfrac{1}{5}-\dfrac{1}{4}\right)-(3-2)\right]$$

$$=4-\left[\left(-\dfrac{1}{5}+\left(-\dfrac{1}{4}\right)\right)-1\right] \quad -\dfrac{1}{5}+\left(-\dfrac{1}{4}\right)=-\dfrac{4}{20}+\left(-\dfrac{5}{20}\right)$$
$$=-\dfrac{9}{20}$$

$$=4-\left[\left(-\dfrac{9}{20}\right)-1\right]$$

$$=4-\left[-\dfrac{9}{20}+(-1)\right]$$

$$=4-\left[-\dfrac{9}{20}+\left(-\dfrac{20}{20}\right)\right]$$

$$=4-\left(-\dfrac{29}{20}\right)=4+\dfrac{29}{20}$$

$$=\dfrac{80}{20}+\dfrac{29}{20}=\dfrac{109}{20}$$

Objective 3 Practice Exercises

For extra help, see Example 3 on page 66 of your text.

Perform each operation.

7. $\left[8-(-12)\right]-2$ 7. _____

8. $3-\left[-4+(11-19)\right]$ 8. _____

9. $\dfrac{2}{9}-\left[\dfrac{5}{6}-\left(-\dfrac{2}{3}\right)\right]$ 9. _____

Objective 4   Translate words and phrases that indicate subtraction.

Video Examples

Review these examples for Objective 4:

4. Write a numerical expression for each phrase, and simplify the expression.

   a. The difference between −10 and 7

$-10-7$ simplifies to $-10+(-7)$, which equals −17

   b. 3 subtracted from the sum of 9 and −4

First, add 9 and −4. Next subtract 3 from this sum.
$\left[9+(-4)\right]-3$ simplifies to $5-3$, which equals 2.

Now Try:

4. Write a numerical expression for each phrase, and simplify the expression.

   a. The difference between −17 and 9

             _____

   b. 8 subtracted from the sum of 25 and −6

             _____

5. The early morning temperature on a mountain in California was –8°F. At noon the temperature was 38°F. What was the rise in temperature?

We must subtract the lowest temperature from the highest temperature.

$$38 - (-8) = 38 + 8 = 46$$

The rise was 46°F.

5. The floor of Death Valley is 282 ft below sea level. A nearby mountain has an elevation of 5182 ft above sea level. Find the difference between the highest and lowest elevations.

_____

Objective 4 Practice Exercises

For extra help, see Examples 4–5 on pages 67–68 of your text.

Write a numerical expression for each phrase, and then simplify the expression.

10. 4 less than –4

10. _____

11. The sum of –4 and 12, decreased by 9

11. _____

Solve the problem.

12. Dr. Somers runs an experiment at –43.3°C. He then lowers the temperature by 7.9°C. What is the new temperature for the experiment?

12. _____

Chapter 1 THE REAL NUMBER SYSTEM

1.6   Multiplying and Dividing Real Numbers

Learning Objectives
1   Find the product of a positive number and a negative number.
2   Find the product of two negative numbers.
3   Use the reciprocal of a number to apply the definition of division.
4   Use the rules for order of operations when multiplying and dividing real numbers.
5   Evaluate expressions involving variables.
6   Translate words and phrases that indicate multiplication and division.
7   Translate simple sentences into equations.

Key Terms

Use the vocabulary terms listed below to complete each statement in exercises 1−5.

product   **quotient**   **dividend**

divisor   **reciprocals**

1. The answer to a division problem is called the _____.

2. Pairs of numbers whose product is 1 are called _____.

3. The answer to a multiplication problem is called the _____.

4. In the division $x \div y$, x is called the _____.

5. In the division $x \div y$, y is called the _____.

Objective 1   Find the product of a positive number and a negative number.

Video Examples

Review these examples for Objective 1:
1. Find each product using the multiplication rule.

   a. $9(-6)$

   $9(-6) = -(9 \cdot 6) = -54$

   c. $-3.7(2.5)$

   $-3.7(2.5) = -9.25$

Now Try:
1. Find each product using the multiplication rule.

   a. $8(-7)$

   _____

   c. $-9.8(4.6)$

   _____

Objective 1 Practice Exercises

For extra help, see Example 1 on page 74 of your text.

Find each product.

   1. $7(-4)$

1. _____

   2. $\left(\frac{1}{5}\right)\left(-\frac{2}{3}\right)$

2. _____

   3. $(-3.2)(4.1)$

3. _____

Objective 2   Find the product of two negative numbers.

Video Examples

Review these examples for Objective 2:

2. Find each product using the multiplication rule.

    a. $-7(-3)$

    $-7(-3) = 21$

    b. $-8(-13)$

    $-8(-13) = 104$

Now Try:

2. Find each product using the multiplication rule.

    a. $-5(-6)$

    _____

    b. $-9(-15)$

    _____

Objective 2 Practice Exercises

For extra help, see Example 2 on page 75 of your text.

Find each product.

   4. $(-4)(-10)$

4. _____

   5. $\left(-\frac{2}{7}\right)\left(-\frac{14}{5}\right)$

5. _____

   6. $(-0.4)(-3.4)$

6. _____

Objective 3   **Use the reciprocal of a number to apply the definition of division.**

Video Examples

Review these examples for Objective 3:

3.   Find each quotient.

    a. $\dfrac{15}{-3}$

    $\dfrac{15}{-3} = -5$

    b. $-\dfrac{30}{6}$

    $-\dfrac{30}{6} = -5$

    c. $\dfrac{-7.5}{-0.03}$

    $\dfrac{-7.5}{-0.03} = 250$

    d. $-\dfrac{1}{9} \div \left(-\dfrac{2}{3}\right)$

    $-\dfrac{1}{9} \div \left(-\dfrac{2}{3}\right) = -\dfrac{1}{9} \cdot \left(-\dfrac{3}{2}\right) = \dfrac{1}{6}$

Now Try:

3.   Find each quotient.

    a. $\dfrac{-18}{-6}$

    _____

    b. $\dfrac{16}{-8}$

    _____

    c. $\dfrac{-18.3}{-6.1}$

    _____

    d. $-\dfrac{2}{5} \div \left(-\dfrac{11}{10}\right)$

    _____

Objective 3 Practice Exercises

For extra help, see Example 3 on page 77 of your text.

Find each quotient.

7. $\dfrac{-120}{-20}$ 7. _____

8. $\dfrac{0}{-2}$ 8. _____

9. $\dfrac{10}{0}$ 9. _____

Name: _____   Date: _____

Instructor: _____   Section: _____

Objective 4   **Use the rules for order of operations when multiplying and dividing real numbers.**

Video Examples

Review this example for Objective 4:	**Now Try:**
4d. Simplify.	**4d.** Simplify.

$$\frac{6(-4)-5(3)}{3(2-7)}$$ $$\frac{-9(-3)+4(-8)}{-4(5-6)}$$

$$\frac{6(-4)-5(3)}{3(2-7)}=\frac{-24-15}{3(-5)}$$

$$=\frac{-39}{-15}$$ _____

$$=\frac{13}{5}$$

Objective 4 Practice Exercises

For extra help, see Example 4 on page 78 of your text.

Perform the indicated operations.

10. $-4\big[(-2)(7)-2\big]$ **10.** _____

11. $\dfrac{-7(2)-(-3)}{5+(-3)}$ **11.** _____

12. $\dfrac{-4\big[8-(-3+7)\big]}{-6\big[3-(-2)\big]-3(-3)}$ **12.** _____

Objective 5   Evaluate expressions involving variables.

Video Examples

Review these examples for Objective 5:

5. Evaluate each expression for $x = -2$, $y = -4$, and $m = -5$.

a. $(5x + 6y)(-3m)$

Substitute the given values for the variables. Then simplify.

$(5x + 6y)(-3m)$

$= [5(-2) + 6(-4)][-3(-5)]$

$= [-10 + (-24)][15]$

$= [-34]15$

$= -510$

b. $4x^2 - 5y^2$

$4x^2 - 5y^2 = 4(-2)^2 - 5(-4)^2$

$\qquad\qquad = 4(4) - 5(16)$

$\qquad\qquad = 16 - 80$

$\qquad\qquad = -64$

Now Try:

5. Evaluate each expression for $x = -5$, $y = -3$, and $p = -4$.

a. $(6x + 2y)(-3p)$

_____

b. $7x^2 - 8y^2$

_____

Objective 5 Practice Exercises

For extra help, see Example 5 on page 79 of your text.

Evaluate the following expressions if x = −3, y = 2, and a = 4.

13. $-x + \left[(-a + y) - 2x\right]$

13. _____

14. $(-4 + x)(-a) - |x|$

14. _____

15. $\dfrac{4a - x}{y^2}$

15. _____

Name: _____   Date: _____
Instructor: _____   Section: _____

Objective 6 Translate words and phrases that indicate multiplication and division.

Video Examples

Review these examples for Objective 6:

6. Write a numerical expression for each phrase, and simplify the expression.

   a. The product of 15 and the sum of 4 and –7

   $15[4 + (-7)]$ simplifies to $15[-3]$, which equals –45.

   d. 45% of the difference between 24 and –5

   $0.45[24 - (-5)]$ simplifies to $0.45[29]$, which equals 13.05.

Now Try:

6. Write a numerical expression for each phrase, and simplify the expression.

   a. The product of 16 and the sum of 5 and –7

   _____

   d. 8% of the difference between 18 and –4

   _____

Objective 6 Practice Exercises

For extra help, see Examples 6–7 on pages 80–81 of your text.

Write a numerical expression for each phrase and simplify.

16. The product of –7 and 3, added to –7

16. _____

17. Three-tenths of the difference between 50 and –10, subtracted from 85

17. _____

18. The sum of –12 and the quotient of 49 and –7

18. _____

Objective 7 Translate simple sentences into equations.

Video Examples

Review this example for Objective 7:

8d. Write the sentence in symbols, using x to represent the number.

   The quotient of 27 and a number is –3.

   $\dfrac{27}{x} = -3$

Now Try:

8d. Write the sentence in symbols, using x to represent the number.

   The quotient of 36 and a number is –4

   _____

Objective 7 Practice Exercises

For extra help, see Example 8 on page 81 of your text.

Write each statement in symbols, using x as the variable.

19.   Two-thirds of a number is –7.   19. _____

20.   –8 times a number is 72.   20. _____

21.   When a number is divided by –4, the result is 1.   21. _____

49

Chapter 1 THE REAL NUMBER SYSTEM

1.7   Properties of Real Numbers

Learning Objectives
1   Use the commutative properties.
2   Use the associative properties.
3   Use the identity properties.
4   Use the inverse properties.
5   Use the distributive property.

Key Terms

Use the vocabulary terms listed below to complete each statement in exercises 1–2.

identity element for addition

identity element for multiplication

1.   When the _____, which is 0, is added to a number, the number is unchanged.

2.   When a number is multiplied by the _____, which is 1, the number is unchanged.

Objective 1   Use the commutative properties.

Video Examples

Review these examples for Objective 1:

1.   Use a commutative property to complete each statement.

   a. $-7+6=6+$ _____

   Using the commutative property of addition,
   $-7+6=6+(-7)$

   b. $(-3)5=$ _____ (-3)

   Using the commutative property of multiplication,
   $(-3)5=5(-3)$

Now Try:

1.   Use a commutative property to complete each statement.

   a. $-12+8=8+$ _____

   _____

   b. $(-4)2=$ _____ (-4)

   _____

Objective 1 Practice Exercises

For extra help, see Example 1 on page 88 of your text.

Complete each statement. Use a commutative property.

1. $y + 4 = $ _____ $+ y$ 1. _____

2. $5(2) = $ _____ (5) 2. _____

3. $-4(4 + z) = $ _____ (-4) 3. _____

Objective 2   Use the associative properties.

Video Examples

Review these examples for Objective 2:	**Now Try:**
2. Use an associative property to complete each statement.	2. Use an associative property to complete each statement.

Review these examples for Objective 2:

2. Use an associative property to complete each statement.

    a. $-5 + (3 + 7) = (-5 + $ _____ $) + 7$

    Using the associative property of addition,
$-5 + (3 + 7) = (-5 + 3) + 7$

    b. $[4 \cdot (-9)] \cdot 2 = 4 \cdot$ _____

    Using the associative property of multiplication,
$[4 \cdot (-9)] \cdot 2 = 4 \cdot [(-9) \cdot 2]$

4. Find each sum or product.

    a. $54 + 21 + 3 + 17 + 29$

    $54 + 21 + 3 + 17 + 29$
    $= 54 + (21 + 29) + (3 + 17)$
    $= 54 + 50 + 20$
    $= 124$

    b. $50(43)(4)$

    $50(43)(4) = 50(4)(43)$
             $= 200(43)$
             $= 8600$

Now Try:

2. Use an associative property to complete each statement.

    a. $-8 + (4 + 6) = (-8 + $ ___ $) + 6$

    _____

    b. $[8 \cdot (-3)] \cdot 4 = 8 \cdot$ _____

    _____

4. Find each sum or product.

    a. $48 + 15 + 12 + 24 + 8$

    _____

    b. $40(63)(5)$

    _____

Objective 2 Practice Exercises

For extra help, see Examples 2–4 on pages 88–89 of your text.

Complete each statement. Use an associative property.

4. $4(ab) = \underline{\qquad} \cdot b$ 4. _____

5. $[x + (-4)] + 3y = x + \underline{\qquad}$ 5. _____

6. $4r + (3s + 14t) = \underline{\qquad} + 14t$ 6. _____

Objective 3   Use the identity properties.

Video Examples

Review this example for Objective 3:   **Now Try:**

6a. Write $\dfrac{56}{35}$ in lowest terms.   **6a.** Write $\dfrac{49}{63}$ in lowest terms.

$$\frac{56}{35} = \frac{8 \cdot 7}{5 \cdot 7}$$ _____

$$= \frac{8}{5} \cdot \frac{7}{7}$$

$$= \frac{8}{5} \cdot 1$$

$$= \frac{8}{5}$$

Objective 3 Practice Exercises

For extra help, see Examples 5–6 on page 90 of your text.

Use an identity property to complete each statement.

7. $4 + 0 = \underline{\qquad}$ 7. _____

8. $\underline{\qquad} \cdot 1 = 12$ 8. _____

Use an identity property to simplify the expression.

9. $\dfrac{30}{35}$ 9. _____

Objective 4   Use the inverse properties.

Video Examples

Review these examples for Objective 4:

7. Use an inverse property to complete each statement.

   b. $5 + \_\_\_\_\_ = 0$

   Use the inverse property of addition.
   $$5 + (-5) = 0$$

   d. $\_\_\_\_\_ \cdot \dfrac{6}{7} = 1$

   Use the inverse property of multiplication.
   $$\dfrac{7}{6} \cdot \dfrac{6}{7} = 1$$

Now Try:

7. Use an inverse property to complete each statement.

   b. $8 + \_\_\_\_\_ = 0$

   _____

   d. $\dfrac{8}{5} \cdot \_\_\_\_\_ = 1$

   _____

Objective 4 Practice Exercises

For extra help, see Example 7 on page 91 of your text.

Complete the statements so that they are examples of either an identity property or an inverse property. Identify which property is used.

10. $-4 + \_\_\_\_\_ = 0$ 10. _____

11. $-9 + \_\_\_\_\_ = -9$ 11. _____

12. $-\dfrac{3}{5} \cdot \_\_\_\_\_ = 1$ 12. _____

Objective 5   Use the distributive property.

Video Examples

Review these examples for Objective 5:

8f. Use the distributive property to rewrite the expression.

   $$4 \cdot 9 + 4 \cdot 2$$

   Use the distributive property in reverse.
   $$\begin{aligned} 4 \cdot 9 + 4 \cdot 2 &= 4(9 + 2) \\ &= 4(11) \\ &= 44 \end{aligned}$$

Now Try:

8f. Use the distributive property to rewrite the expression.

   $$25 \cdot 9 + 25 \cdot 6$$

   _____

9c. Write the expression without parentheses.

$$-(-p-5r+9x)$$

$$-(-p-5r+9x)$$
$$=-1\cdot(-1p-5r+9x)$$
$$=-1\cdot(-1p)-1\cdot(-5r)-1\cdot(9x)$$
$$=p+5r-9x$$

9c. Write the expression without parentheses.
$$-(-4x-5y+z)$$

_____

Objective 5 Practice Exercises

For extra help, see Examples 8–9 on pages 92–93 of your text.

Use the distributive property to rewrite each expression. Simplify if possible.

13. $n(2a-4b+6c)$

13. _____

14. $-2(5y-9z)$

14. _____

15. $-(-2k+7)$

15. _____

Chapter 1 THE REAL NUMBER SYSTEM

1.8   Simplifying Expressions

Learning Objectives
1   Simplify expressions.
2   Identify terms and numerical coefficients.
3   Identify like terms.
4   Combine like terms.
5   Simplify expressions from word phrases.

Key Terms

Use the vocabulary terms listed below to complete each statement in exercises 1−3.

term   numerical coefficient   like terms

1.   In the term $4x^2$, "4" is the_____.

2.   A number, a variable, or a product or quotient of a number and one or more variables raised to powers is called a _____.

3.   Terms with exactly the same variables, including the same exponents, are called

_____.

Objective 1   Simplify expressions.

Video Examples

Review these examples for Objective 1:

1.   Simplify each expression.

   c. $7+5(9k+6)$

   $$7+5(9k+6)=7+5(9k)+5(6)$$
   $$=7+45k+30$$
   $$=37+45k$$

   d. $9-(4y-6)$

   $$9-(4y-6)=9-1(4y-6)$$
   $$=9-4y+6$$
   $$=15-4y$$

Now Try:

1.   Simplify each expression.

   c. $8+9(2x+7)$

   _____

   d. $8-(7x-3)$

   _____

Objective 1 Practice Exercises

For extra help, see Example 1 on page 98 of your text.

Simplify each expression.

   1. $4(2x+5)+7$ **1.** _____

   2. $-4+s-(12-21)$ **2.** _____

   3. $-2(-5x+2)+7$ **3.** _____

Objective 2   **Identify terms and numerical coefficients.**

Objective 2 Practice Exercises

For extra help, see pages 98–99 of your text.

Give the numerical coefficient of each term.

   4. $-2y^2$ **4.** _____

   5. $\dfrac{7x}{9}$ **5.** _____

   6. $5.6r^5$ **6.** _____

Objective 3   **Identify like terms.**

Objective 3 Practice Exercises

For extra help, see page 99 of your text.

Identify each group of terms as **like** *or* **unlike***.*

   7. $4x^2, -7x^2$ **7.** _____

   8. $-8m, -8m^2$ **8.** _____

   9. $7xy, -6xy^2$ **9.** _____

Objective 4   Combine like terms.

Video Examples

Review these examples for Objective 4:

2c. Combine like terms in the expression.

$$9x + x$$

$$9x + x = 9x + 1x$$
$$= (9 + 1)x$$
$$= 10x$$

3a. Simplify the expression.

$$15y + 3(5 + 4y)$$

$$15y + 3(5 + 4y) = 15y + 3(5) + 3(4y)$$
$$= 15y + 15 + 12y$$
$$= 27y + 15$$

Now Try:

2c. Combine like terms in the expression.
$$18x + x$$

_____

3a. Simplify each expression.

$$9y + 5(3 + 8y)$$

_____

Objective 4 Practice Exercises

For extra help, see Examples 2–3 on pages 99–101 of your text.

Simplify.

10. $12y - 7y^2 + 4y - 3y^2$

10. _____

11. $-4(x + 4) + 2(3x + 1)$

11. _____

12. $2.5(3y + 1) - 4.5(2y - 3)$

12. _____

Name: _____   Date: _____
Instructor: _____   Section: _____

Objective 5   Simplify expressions from word phrases.

Video Examples

Review this example for Objective 5:

4. Translate the phrase into a mathematical expression and simplify.

The sum of 8, three times a number, nine times a number, and seven times a number.

Use x for the number.

$8 + 3x + 9x + 7x$ simplifies to $8 + 19x$.

Now Try:

4. Translate the phrase into a mathematical expression and simplify.

The sum of 11, ten times a number, eight times a number, and four times a number

_____

Objective 5 Practice Exercises

For extra help, see Example 4 on page 101 of your text.

Write each phrase as a mathematical expression and simplify by combining like terms. Use x as the variable.

13. The sum of six times a number and 12, added to four times the number.

13. _____

14. The sum of seven times a number and 2, subtracted from three times the number.

14. _____

15. Four times the difference between twice a number and six times the number, added to six times the sum of the number and 9.

15. _____

Chapter 2 EQUATIONS, INEQUALITIES, AND APPLICATIONS

2.1   The Addition Property of Equality

Learning Objectives

1   Identify linear equations.
2   Use the addition property of equality.
3   Simplify, and then use the addition property of equality.

Key Terms

Use the vocabulary terms listed below to complete each statement in exercises 1−3.

        linear equation   **solution set**   **equivalent equations**

1.   Equations that have exactly the same solutions sets are called

      _____ .

2.   An equation that can be written in the form $Ax + B = C$, where A, B, and C are real numbers and $A \neq 0$, is called a _____ .

3.   The set of all numbers that satisfy an equation is called its _____ .

Objective 1   Identify linear equations.

Objective 1 Practice Exercises

For extra help, see page 118 of your text.

Tell whether each of the following is a linear equation.

1. $3x^2 + 4x + 3 = 0$ 1. _____

2. $\dfrac{5}{x} - \dfrac{3}{2} = 0$ 2. _____

3. $4x - 2 = 12x + 9$ 3. _____

Objective 2   Use the addition property of equality.

Video Examples

Review these examples for Objective 2:

1. Solve $x - 15 = 8$.

$$x - 15 = 8$$
$$x - 15 + 15 = 8 + 15$$
$$x = 23$$

Check $x - 15 = 8$
$$23 - 15 \overset{?}{=} 8$$
$$8 = 8 \quad \text{True}$$

The solution is 23, and the solution set is {23}.

3. Solve $-5 = x + 17$.

$$-5 = x + 17$$
$$-5 - 17 = x + 17 - 17$$
$$-22 = x$$

Check $-5 = x + 17$
$$-5 \overset{?}{=} -22 + 17$$
$$-5 = -5 \quad \text{True}$$

The solution set is {–22}.

4. Solve $-5p = -6p + 3$.

$$-5p = -6p + 3$$
$$-5p + 6p = -6p + 3 + 6p$$
$$p = 3$$

Check by substituting 3 in the original equation.
The solution set is {3}.

Now Try:

1. Solve $x - 12 = 9$.

_____

3. Solve $-10 = x + 9$.

_____

4. Solve $-8p = -9p + 7$.

_____

Objective 2 Practice Exercises

For extra help, see Examples 1–6 on pages 119–121 of your text.

Solve each equation by using the addition property of equality. Check each solution.

4. $y - 4 = 16$

4. _____

5. $\frac{9}{8}p - \frac{1}{2} = \frac{1}{8}p$

5. _____

6. $9.5y - 2.4 = 10.5y$

6. _____

Objective 3   Simplify, and then use the addition property of equality.

Video Examples

Review these examples for Objective 3:

7.   Solve $5t - 16 + t + 4 = 9 + 5t + 6$.

$$5t - 16 + t + 4 = 9 + 5t + 6$$
$$6t - 12 = 15 + 5t$$
$$6t - 12 - 5t = 15 + 5t - 5t$$
$$t - 12 = 15$$
$$t - 12 + 12 = 15 + 12$$
$$t = 27$$

Check by substituting 27 in the original equation. The solution set is {27}.

8.   Solve $4(3 + 6x) - (5 + 23x) = 19$.

$$4(3 + 6x) - (5 + 23x) = 19$$
$$4(3) + 4(6x) - 1(5) - 1(23x) = 19$$
$$12 + 24x - 5 - 23x = 19$$
$$x + 7 = 19$$
$$x + 7 - 7 = 19 - 7$$
$$x = 12$$

Check by substituting 12 in the original equation. The solution set is {12}.

Now Try:

7.   Solve
$$8t - 9 + t + 7 = 12 + 8t + 15.$$

_____

8.   Solve
$$5(7 + 8x) - (29 + 39x) = 14.$$

_____

Objective 3 Practice Exercises

For extra help, see Examples 7–8 on page 122 of your text.

Solve each equation. First simplify each side of the equation as much as possible. Check each solution.

7. $3(t+3)-(2t+7)=9$

7. _____

8. $-4(5g-7)+3(8g-3)=15-4+3g$

8. _____

9. $3.6p+4.8+4.0p=8.6p-3.1+0.7$

9. _____

Chapter 2 EQUATIONS, INEQUALITIES, AND APPLICATIONS

2.2   The Multiplication Property of Equality

Learning Objectives
1   Use the multiplication property of equality.
2   Simplify, and then use the multiplication property of equality.

Key Terms

Use the vocabulary terms listed below to complete each statement in exercises 1−2.

multiplication property of equality   **addition property of equality**

1.   The _____ states that multiplying both sides of an equation by the same nonzero number will not change the solution.

2.   When the same quantity is added to both sides of an equation, the _____ is being applied.

Objective 1   Use the multiplication property of equality.

Video Examples

Review these example for Objective 1:

1.   Solve $6x = 78$.

$$6x = 78$$

$$\frac{6x}{6} = \frac{78}{6}$$

$$x = 13$$

Check $6x = 78$

$$6(13) \overset{?}{=} 78$$

$$78 = 78 \quad \text{True}$$

The solution set is $\{13\}$.

6.   Solve $-k = -22$.

$$-k = -22$$

$$-1 \cdot k = -22$$

$$-1(-1 \cdot k) = -1(-22)$$

$$[-1(-1)] \cdot k = 22$$

$$1 \cdot k = 22$$

$$k = 22$$

Check by substituting 22 in the original equation. The solution set is $\{22\}$.

Now Try:

1.   Solve $4x = 56$.

         _____

6.   Solve $-y = -39$.

         _____

4. Solve $\frac{x}{7} = 5$.

$$\frac{x}{7} = 5$$

$$\frac{1}{7}x = 5$$

$$7 \cdot \frac{1}{7}x = 7 \cdot 5$$

$$x = 35$$

Check by substituting 35 in the original equation. The solution set is {35}.

5. Solve $\frac{5}{6}x = 15$.

$$\frac{5}{6}x = 15$$

$$\frac{6}{5} \cdot \frac{5}{6}x = \frac{6}{5} \cdot 15$$

$$1 \cdot x = \frac{6}{5} \cdot \frac{15}{1}$$

$$x = 18$$

Check by substituting 18 in the original equation. The solution set is {18}.

4. Solve $\frac{x}{8} = 3$.

_____

5. Solve $\frac{7}{9}h = 28$.

_____

Objective 1 Practice Exercises

For extra help, see Examples 1–6 on pages 127–129 of your text.

Solve each equation and check your solution.

1. $-3w = 51$

1. _____

2. $\frac{3p}{7} = -6$

2. _____

3. $-2.7v = -17.28$ **3.** _____

Objective 2   Simplify, and then use the multiplication property of equality.

Video Examples

Review this example for Objective 2:

7.   Solve $9m + 4m = 39$.

$$9m + 4m = 39$$
$$13m = 39$$
$$\frac{13m}{13} = \frac{39}{13}$$
$$m = 3$$

Check by substituting 3 in the original equation.
The solution set is $\{3\}$.

Now Try:

7.   Solve $12m + 8m = 80$.

_____

Objective 2 Practice Exercises

For extra help, see Examples 7–8 on pages 129–130 of your text.

Solve each equation and check your solution.

4. $-7b + 12b = 125$ **4.** _____

5. $3w - 7w = 20$ **5.** _____

6. $-11h - 6h + 14h = -21$ **6.** _____

Chapter 2 EQUATIONS, INEQUALITIES, AND APPLICATIONS

2.3   More on Solving Linear Equations

Learning Objectives
1   Learn and use the four steps for solving a linear equation.
2   Solve equations that have no solution or infinitely many solutions.
3   Solve equations with fractions or decimals as coefficients.
4   Write expressions for two related unknown quantities.

Key Terms

Use the vocabulary terms listed below to complete each statement in exercises 1−3.

conditional equation   identity   contradiction

1.   An equation with no solution is called a(n) _____.

2.   A(n) _____ is an equation that is true for some values of the variable and false for other values.

3.   An equation that is true for all values of the variable is called a(n) _____.

Objective 1   Learn and use the four steps for solving a linear equation.

Video Examples

Review this example for Objective 1:	Now Try:
3. Solve $5(k-4)-k=k-2$.	**3.** Solve $9(k-2)-k=k+10$.

Step 1 Clear parentheses using the distributive property.

$$5(k-4)-k=k-2$$
$$5(k)+5(-4)-k=k-2$$
$$5k-20-k=k-2$$
$$4k-20=k-2$$

Step 2 $4k-20-k=k-2-k$
$$3k-20=-2$$
$$3k-20+20=-2+20$$
$$3k=18$$

Step 3 $\dfrac{3k}{3}=\dfrac{18}{3}$
$$k=6$$

Step 4 Check by substituting 6 for k in the original equation.

$$5(k-4)-k = k-2$$
$$5(6-4)-6 \overset{?}{=} 6-2$$
$$5(2)-6 \overset{?}{=} 4$$
$$10-6 \overset{?}{=} 4$$
$$4 = 4 \quad \text{True}$$

The solution, 6, checks, so the solution set is {6}.

Objective 1 Practice Exercises

For extra help, see Examples 1–5 on pages 133–136 of your text.

Solve each equation and check your solution.

1. $7t+6 = 11t-4$ 1. _____

2. $3a-6a+4(a-4) = -2(a+2)$ 2. _____

3. $3(t+5) = 6-2(t-4)$ 3. _____

Objective 2   Solve equations that have no solution or infinitely many solutions.

Video Examples

Review these examples for Objective 2:	**Now Try:**
6. Solve $6x - 18 = 6(x-3)$.	**6.** Solve $3x + 4(x-5) = 7x - 20$.

$$6x - 18 = 6(x-3)$$
$$6x - 18 = 6x - 18$$
$$6x - 18 - 6x = 6x - 18 - 6x$$
$$-18 = -18$$
$$-18 + 18 = -18 + 18$$
$$0 = 0$$

The solution set is {all real numbers}.

_____

7. Solve $3x + 4(x-5) = 7x + 5$.

$$3x + 4(x-5) = 7x + 5$$
$$3x + 4x - 20 = 7x + 5$$
$$7x - 20 = 7x + 5$$
$$7x - 20 - 7x = 7x + 5 - 7x$$
$$-20 = 5 \quad \text{False}$$

There is no solution. The solution set is $\varnothing$.

7. Solve $-5x + 17 = x - 6(x+3)$.

_____

Objective 2 Practice Exercises

For extra help, see Examples 6–7 on page 137 of your text.

Solve each equation and check your solution.

4. $3(6x - 7) = 2(9x - 6)$

4. _____

5. $6y - 3(y+2) = 3(y-2)$

5. _____

6. $3(r-2)-r+4=2r+6$ **6.** _____

Objective 3   **Solve equations with fractions or decimals as coefficients.**

Video Examples

Review these examples for Objective 3:   **Now Try:**

8.   Solve $\frac{3}{4}x-\frac{1}{2}x=-\frac{1}{8}x-6$.   **8.**   Solve $\frac{2}{9}x-\frac{1}{6}x=\frac{2}{3}x+11$.

Multiply each side by 8, the LCD.

$$\frac{3}{4}x-\frac{1}{2}x=-\frac{1}{8}x-6$$

Step 1 $8\left(\frac{3}{4}x-\frac{1}{2}x\right)=8\left(-\frac{1}{8}x-6\right)$

$$8\left(\frac{3}{4}x\right)+8\left(-\frac{1}{2}x\right)=8\left(-\frac{1}{8}x\right)+8(-6)$$

$$6x-4x=-x-48$$

$$2x=-x-48$$

Step 2 $2x+x=-x-48+x$

$$3x=-48$$

Step 3 $\frac{3x}{3}=-\frac{48}{3}$

$$x=-16$$

Step 4 $\frac{3}{4}x-\frac{1}{2}x=-\frac{1}{8}x-6$

$$\frac{3}{4}(-16)-\frac{1}{2}(-16)\overset{?}{=}-\frac{1}{8}(-16)-6$$

$$-12+8\overset{?}{=}2-6$$

$$-4=-4 \quad \text{True}$$

The solution set is $\{-16\}$.

10. Solve $0.2x + 0.04(10 - x) = 0.06(4)$.

To clear decimals, multiply by 100.
$$0.2x + 0.04(10 - x) = 0.06(4)$$
Step 1 $\quad 100[0.2x + 0.04(10 - x)] = 100[0.06(4)]$
$$100(0.2x) + 100[0.04(10 - x)] = 100[0.06(4)]$$
$$20x + 4(10) + 4(-x) = 24$$
$$20x + 40 - 4x = 24$$
$$16x + 40 = 24$$
Step 2 $\qquad 16x + 40 - 40 = 24 - 40$
$$16x = -16$$
Step 3 $\qquad \dfrac{16x}{16} = \dfrac{-16}{16}$
$$x = -1$$
Step 4 Check to confirm that $\{-1\}$ is the solution set.

10. Solve
$$0.5x + 0.04(5 - 8x) = 0.07(8).$$

_____

Objective 3 Practice Exercises

For extra help, see Examples 8–10 on pages 138–139 of your text.

Solve each equation and check your solution.

7. $\quad \dfrac{3}{8}x - \dfrac{1}{3}x = \dfrac{1}{12}$

7. _____

8. $\quad \dfrac{1}{3}(2m - 1) - \dfrac{3}{4}m = \dfrac{5}{6}$

8. _____

9. $0.45a - 0.35(20 - a) = 0.02(50)$ 9. _____

Objective 4   **Write expressions for two related unknown quantities.**

Video Examples

Review this example for Objective 4:

11a. Perform the translation.

Two numbers have a sum of 51. If one of the numbers is represented by x, find an expression for the other number.

If one number is x, then the other number is obtained by subtracting x from 51.

    $51 - x$.

To check, we find the sum of the two numbers.

    $x + (51 - x) = 51$

Now Try:

11a. Perform the translation.

Two numbers have a sum of 67. If one of the numbers is represented by t, find an expression for the other number.

         _____

Objective 4 Practice Exercises

For extra help, see Example 11 on page 140 of your text.

Write an expression for the two related unknown quantities.

10.   Two numbers have a sum of 36. One is m. Find the   10. _____
        other number.

11.   The product of two numbers is 17. One number is p.   **11.** _____
What is the other number?

12.   Admission to the circus costs x dollars for an adult   **12.** _____
and y dollars for a child. Find the total cost of 6
adults and 4 children.

Chapter 2 EQUATIONS, INEQUALITIES, AND APPLICATIONS

2.4   An Introduction to Applications of Linear Equations

Learning Objectives	
1	Learn the six steps for solving applied problems.
2	Solve problems involving unknown numbers.
3	Solve problems involving sums of quantities.
4	Solve problems involving consecutive integers.
5	Solve problems involving complementary and supplementary angles.

Key Terms

Use the vocabulary terms listed below to complete each statement in exercises 1−5.

> **complementary angles**   **right angle**   **supplementary angles**
>
> **straight angle**   **consecutive integers**

1.   Two angles whose measures sum to 180° are _____.

2.   Two angles whose measures sum to 90° are _____.

3.   An angle whose measure is exactly 90° is a _____.

4.   An angle whose measure is exactly 180° is a _____.

5.   Two integers that differ by 1 are _____.

Objective 1   Learn the six steps for solving applied problems.

Objective 1 Practice Exercises

For extra help, see page 147 of your text.

1.   Write the six problem-solving steps.   1. _____

                                                           _____

                                                             _____

                                                             _____

                                                             _____

                                                             _____

Objective 2   Solve problems involving unknown numbers.

Video Examples

Review this example for Objective 2:

1. The product of 5, and a number decreased by 8, is 150. What is the number?

    Step 1 Read the problem carefully. We are asked to find a number.

    Step 2 Assign a variable to represent the unknown quantity.
        Let x = the number.

    Step 3 Write an equation.
    The product   a   decreased
        of 5,   and number   by   8, is 150.
         $\downarrow$ $\downarrow$ $\downarrow$ $\downarrow$ $\downarrow$ $\downarrow$
         $5\cdot$ $(x$ $-$ $8) = 150$

    Step 4 Solve the equation.
    $$5(x-8) = 150$$
    $$5x - 40 = 150$$
    $$5x - 40 + 40 = 150 + 40$$
    $$5x = 190$$
    $$\frac{5x}{5} = \frac{190}{5}$$
    $$x = 38$$

    Step 5 State the answer. The number is 38.

    Step 6 Check. The number 38 decreased by 8 is 30. The product of 5 and 30 is 150. The answer, 38, is correct.

Now Try:

1. The product of 8, and a number decreased by 11, is 40. What is the number?

    _____

Objective 2 Practice Exercises

For extra help, see Example 1 on page 147 of your text.

Write an equation for each of the following and then solve the problem. Use x as the variable.

2. If 4 is added to 3 times a number, the result is 7. Find the number.

    2. _____

    _____

3. If −2 is multiplied by the difference between 4 and a number, the result is 24. Find the number.

3. _____

_____

4. If four times a number is added to 7, the result is five less than six times the number. Find the number.

4. _____

_____

Objective 3   Solve problems involving sums of quantities.

Video Examples

Review this example for Objective 3:

2. George and Al were opposing candidates in the school board election. George received 21 more votes than Al, with 439 votes cast. How many votes did Al receive?

Step 1 Read the problem carefully. We are given total votes and asked to find the number of votes Al received.

Step 2 Assign a variable.
Let x = the number of votes Al received.
Then $x + 21$ = the number of votes George received.

Step 3 Write an equation.

The total	is	votes for Al	plus	votes for George
↓	↓	↓	↓	↓
439	=	x	+	$(x+21)$

Now Try:

2. On a psychology test, the highest grade was 38 points more than the lowest grade. The sum of the two grades was 142. Find the lowest grade.

_____

Step 4 Solve the equation.

$$439 = x + (x + 21)$$
$$439 = 2x + 21$$
$$439 - 21 = 2x + 21 - 21$$
$$418 = 2x$$
$$\frac{418}{2} = \frac{2x}{2}$$
$$209 = x \quad \text{or} \quad x = 209$$

Step 5 State the answer. Al received 209 votes.

Step 6 Check. George won $209 + 21 = 230$ votes. The total number of votes is $209 + 230 = 439$. The answer checks.

Objective 3 Practice Exercises

For extra help, see Examples 2–4 on pages 148–150 of your text.

Write an equation for each of the following and then solve the problem. Use x as the variable.

5. Mount McKinley in Alaska is 5910 feet higher than Mount Rainier in Washington. Together, their heights total 34,730 feet. How high is each mountain?

    5. _____

    Mt. Rainier _____

    Mt. McKinley _____

6. Charles bought five general admission tickets and four student tickets for a movie. He paid $35.25. If each student ticket cost $3.50, how much did each general admission ticket cost?

    6. _____

    _____

7.   Pablo, Faustino, and Mark swim at a public pool each day for exercise. One day Pablo swam five more than three times as many laps as Mark, and Faustino swam four times as many laps as Mark. If the men swam 29 laps altogether, how many laps did each one swim?

7. _____

Mark _____

Pablo _____

Faustino _____

Objective 4   Solve problems involving consecutive integers.

Video Examples

Review these examples for Objective 4:

5.   Two pages that face each other in this book have 337 as the sum of their page numbers. What are the page numbers?

Step 1   Read the problem. Because the two pages face each other, they must have page numbers that are consecutive integers.

Step 2   Assign a variable.
    Let x = the lesser page number.
Then $x + 1$ = the greater page number.

Step 3   Write an equation. The sum of the page numbers is 337.

$$x + (x + 1) = 337$$

Step 4   Solve the equation.

$$2x + 1 = 337$$
$$2x = 336$$
$$x = 168$$

Step 5   State the answer. The lesser page number is 168, and the greater is $168 + 1 = 169$.

Step 6   Check. The sum of 168 and 169 is 337. The answer is correct.

Now Try:

5.   Two pages that face each other in this book have 705 as the sum of their page numbers. What are the page numbers?

_____

6. Find two consecutive odd integers such that if three times the smaller is added to twice the larger, the sum is 69.

Step 1 Read the problem. We must find two consecutive odd integers.

Step 2 Assign a variable.
   Let x = the lesser consecutive odd integer.
Then $x + 2$ = the greater consecutive odd integer.

Step 3 Write an equation.

Three times the smaller	is added to	twice the larger	is	69.
↓	↓	↓	↓	↓
$3x$	$+$	$2(x+2)$	$=$	69

Step 4 Solve the equation.
$$3x + 2x + 4 = 69$$
$$5x + 4 = 69$$
$$5x = 65$$
$$x = 13$$

Step 5 State the answer. The lesser integer is 13. The greater is $13 + 2 = 15$.

Step 6 Check. Three times the smaller is 39, added to twice the larger, 30, is a sum of 69. The answers check.

6. The sum of four consecutive even integers is 4. Find the integers.

_____

Objective 4 Practice Exercises

For extra help, see Examples 5–6 on pages 151–152 of your text.

Solve each problem.

8. Find two consecutive even integers such that the smaller, added to twice the larger, is 292.

8. _____

9.   Find two consecutive integers such that the larger, added to three times the smaller, is 109.

9. _____

10.   Find three consecutive odd integers whose sum is 363.

10. _____

Objective 5   **Solve problems involving complementary and supplementary angles.**

Video Examples

Review this example for Objective 5:

7.   Find the measure of an angle such that the difference between the measures of an angle and its complement is 20°.

Step 1 Read the problem. We must find the measure of an angle.

Step 2 Assign a variable.

    Let x = the degree measure of the angle
Then $90 - x$ = the degree measure of its complement.

Step 3 Write an equation.

The angle	minus	Measure of the complement	is	20
↓	↓	↓	↓	↓
x	$-$	$(90 - x)$	$=$	20

Now Try:

7.   Find the measure of an angle whose complement is 4 times its measure.

_____

Step 4 Solve the equation.

$$x - 90 + x = 20$$

$$2x - 90 = 20$$

$$2x = 110 \qquad \textit{Step 5} \text{ State the}$$

$$\frac{2x}{2} = \frac{110}{2}$$

$$x = 55$$

answer. The angle is 55°.

Step 6 Check. If the angle measures 55°, then its complement measures $90° - 55° = 35°$. The difference between 55° and 35° is 20°. The answer is correct.

Objective 5 Practice Exercises

For extra help, see Examples 7–8 on pages 153–154 of your text.

Solve each problem.

11.   Find the measure of an angle if the measure of the angle is 8° less than three times the measure of its supplement.

11. _____

12.   Find the measure of an angle whose supplement measures 20° more than twice its complement.

12. _____

13.   Find the measure of an angle whose complement is   **13.** _____
       9° more than twice its measure.

Chapter 2 EQUATIONS, INEQUALITIES, AND APPLICATIONS

2.5   Formulas and Additional Applications from Geometry

Learning Objectives
1   Solve a formula for one variable, given the values of the other variables.
2   Use a formula to solve an applied problem.
3   Solve problems involving vertical angles and straight angles.
4   Solve a formula for a specified variable.

Key Terms

Use the vocabulary terms listed below to complete each statement in exercises 1−4.

formula   area   perimeter   vertical angles

1. The nonadjacent angles formed by two intersecting lines are called

_____.

2. An equation in which variables are used to describe a relationship is called a(n)

_____.

3. The distance around a figure is called its _____.

4. A measure of the surface covered by a figure is called its _____.

Objective 1   Solve a formula for one variable, given the values of the other variables.

Video Examples

Review this example for Objective 1:	Now Try:
1a. Find the value of the remaining variable in each formula.	**1a.** Find the value of the remaining variable in each formula.
$A = LW;\;\; A = 54, L = 8$	$A = LW;\;\; A = 88, L = 16$
Substitute the given values for A and L into the formula.	
$A = LW$	_____
$54 = 8W$	
$\dfrac{54}{8} = \dfrac{8W}{8}$	
$6.75 = W$	
The width is 6.75. Since $8(6.75) = 54$, the answer checks.	

Objective 1 Practice Exercises

For extra help, see Example 1 on page 161 of your text.

In the following exercises, a formula is given, along with the values of all but one of the variables in the formula. Find the value of the variable that is not given.

1. $S = \dfrac{a}{1-r}; S = 60, r = 0.4$

1. _____

2. $I = prt; I = 288, r = 0.04, t = 3$

2. _____

3. $A = \frac{1}{2}(b + B)h; b = 6, B = 16, A = 132$

3. _____

Objective 2   Use a formula to solve an applied problem.

Video Examples

Review these examples for Objective 2:

2. Find the dimensions of a rectangle. The length is 4 m less than three times the width. The perimeter is 96 m.

Step 1 Read the problem. We must find the dimensions of the rectangle.

Step 2 Assign a variable.
Let W = the width of the rectangle, in meters.
Then $L = 3W - 4$ is the length, in meters.

Step 3 Write an equation. Use the formula for the perimeter of a rectangle. Substitute $3W - 4$ for the length.

$$P = 2L + 2W$$

$$96 = 2(3W - 4) + 2W$$

Now Try:

2. Ruth has 42 feet of binding for a rectangular rug that she is weaving. If the rug is 9 feet wide, how long can she make the rug if she wishes to use all the binding on the perimeter of the rug?

_____

Step 4 Solve.
$$96 = 6W - 8 + 2W$$
$$96 = 8W - 8$$
$$96 + 8 = 8W - 8 + 8$$
$$104 = 8W$$
$$\frac{104}{8} = \frac{8W}{8}$$
$$13 = W$$

Step 5 State the answer. The width is 13 m. The length is $3(13) - 4 = 35$ m.

Step 6 Check. The perimeter is $2(13) + 2(35) = 96$ m. The answer checks.

3. The longest side of a triangle is 4 feet longer than the shortest side. The medium side is 2 feet longer than the shortest side. If the perimeter is 36 feet, what are the lengths of the three sides?

Step 1 Read the problem. We must find the lengths of the sides.

Step 2 Assign a variable.
Let $s =$ the length of the shortest side, in feet.
Then $s + 2 =$ the length of the medium side, in feet,
and $s + 4 =$ the length of the longest side, in feet.

Step 3 Write an equation. Use the formula for the perimeter of a triangle.
$$P = a + b + c$$
$$36 = s + (s + 2) + (s + 4)$$

Step 4 Solve.
$$36 = 3s + 6$$
$$30 = 3s$$
$$10 = s$$

Step 5 State the answer. Since s represents the length of the shortest side, its measure is 10 ft.
$s + 2 = 10 + 2 = 12$ ft is the length of the medium side.
$s + 4 = 10 + 4 = 14$ ft is the length of the longest side.

Step 6 Check. The perimeter is $10 + 12 + 14 = 36$ ft, as required.

3. The longest side of a triangle is twice as long as the shortest side. The medium side is 5 feet longer than the shortest side. If the perimeter is 65 feet, what are the lengths of the three sides?

_____

Name: _____   Date: _____

Instructor: _____   Section: _____

Objective 2 Practice Exercises

For extra help, see Examples 2–4 on pages 162–163 of your text.

Use a formula to write an equation for each of the following applications; then solve the application. (Use 3.14 as an approximation for π.)

4. Find the height of a triangular banner whose area is 48 square inches and base is 12 inches.

    4. _____

5. Linda invests $5000 at 6% simple interest and earns $450. How long did Linda invest her money?

    5. _____

6. The circumference of a circular garden is 628 feet. Find the area of the garden. (Hint: First find the radius of the garden.)

    6. _____

Name: Date:

Instructor: Section:

Objective 3   Solve problems involving vertical angles and straight angles.

Video Examples

Review this example for Objective 3:

5b. Find the measure of the marked angles in the figure below.

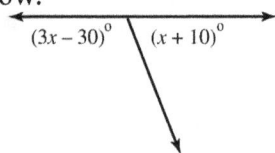

The measures of the marked angles must add to 180° because together they form a straight angle. The angles are supplements of each other.

$$(3x - 30) + (x + 10) = 180$$
$$4x - 20 = 180$$
$$4x = 200$$
$$x = 50$$

Replace x with 50 in the measure of each marked angle.

$$3x - 30 = 3(50) - 30 = 150 - 30 = 120$$
$$x + 10 = 50 + 10 = 60$$

The two angles measure 120° and 60°.

Now Try:

5b. Find the measure of the marked angles in the figure below.

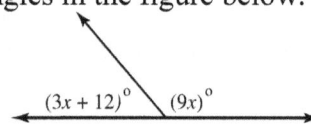

_____

Objective 3 Practice Exercises

For extra help, see Example 5 on page 164 of your text.

Find the measure of each marked angle.

7.

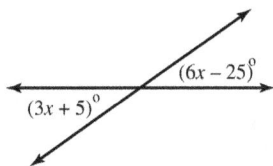

7. _____

8.

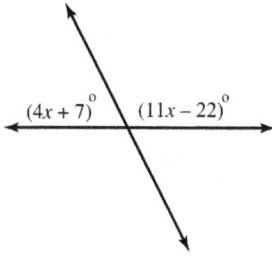

$(4x + 7)^{\circ}$ $(11x - 22)^{\circ}$

9.

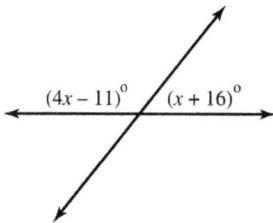

$(4x - 11)^{\circ}$ $(x + 16)^{\circ}$

9. _____

Objective 4   Solve a formula for a specified variable.

Video Examples

Review these examples for Objective 4:

6.   Solve $A = \frac{1}{2}bh$ for h.

$$A = \frac{1}{2}bh$$

$$2A = bh$$

$$\frac{2A}{b} = \frac{bh}{b}$$

$$\frac{2A}{b} = h \quad \text{or} \quad h = \frac{2A}{b}$$

7.   Solve $A = p + prt$ for r.

$$A = p + prt$$

$$A - p = p + prt - p$$

$$A - p = prt$$

$$\frac{A-p}{pt} = \frac{prt}{pt}$$

$$\frac{A-p}{pt} = r \quad \text{or} \quad r = \frac{A-p}{pt}$$

9b. Solve the equation for y.

$$-4x + 5y = 15$$

$$-4x + 5y = 15$$

$$-4x + 5y + 4x = 15 + 4x$$

$$5y = 4x + 15$$

$$\frac{5y}{5} = \frac{4x+15}{5}$$

$$y = \frac{4x}{5} + \frac{15}{5}$$

$$y = \frac{4}{5}x + 3$$

Now Try:

6.   Solve $d = rt$ for t.

_____

7.   Solve $P = a + b + c$ for a.

_____

9b. Solve the equation for y.

$$-18x + 3y = 15$$

_____

Objective 4 Practice Exercises

For extra help, see Examples 6–9 on pages 165–166 of your text.

Solve each formula for the specified variable.

10. $V = LWH$ for H

10. _____

11. $S = (n-2)180$ for n

11. _____

12. $V = \frac{1}{3}\pi r^2 h$ for h

12. _____

Chapter 2 EQUATIONS, INEQUALITIES, AND APPLICATIONS

2.6   Ratio, Proportion, and Percent

Learning Objectives
1   Write ratios.
2   Solve proportions.
3   Solve applied problems using proportions.
4   Find percents and percentages.

Key Terms

Use the vocabulary terms listed below to complete each statement in exercises 1−4.

     ratio   proportion   cross products   terms

1.   A _____ is a statement that two ratios are equal.

2.   A _____ is a comparison of two quantities using a quotient.

3.   In the proportion, $\frac{a}{b} = \frac{c}{d}$, a, b, c, and d are called the _____.

4.   To see whether a proportion is true, determine if the _____ are equal.

Objective 1   Write ratios.

Video Examples

Review these examples for Objective 1:
1.   Write a ratio for each word phrase.

    a. 7 hr to 9 hr

$$\frac{7 \text{ hr}}{9 \text{ hr}} = \frac{7}{9}$$

    b. 15 hr to 4 days

First convert 4 days to hours.
    4 days $= 4 \cdot 24 = 96$ hr
Now write the ratio using the common unit of measure, hours.

$$\frac{15 \text{ hr}}{4 \text{ days}} = \frac{15 \text{ hr}}{96 \text{ hr}} = \frac{15}{96}, \quad \text{or} \quad \frac{5}{32}$$

Now Try:
1.   Write a ratio for each word phrase.

    a. 11 hr to 17 hr

    _____

    b. 32 hr to 5 days

    _____

2. The local grocery store charges the following prices for a bottle of olive oil.

> 16-ounce bottle: $6.99
> 25.5-ounce bottle: $9.99
> 32-ounce bottle: $12.99
> 44-ounce bottle: $14.99

Which size is the best buy? That is, which size has the lowest unit price?

To find the best buy, write ratios comparing the price for each size bottle to the number of units (ounces) per bottle.

Size	Unit price (dollars per ounce)
16 oz	$\frac{\$6.99}{16} = \$0.437$
25.5 oz	$\frac{\$9.99}{25.5} = \$0.392$
32 oz	$\frac{\$12.99}{32} = \$0.406$
44 oz	$\frac{\$14.99}{44} = \$0.341$

Because the 44-oz size has the lowest unit price, $0.341, it is the best buy.

2. The local grocery store charges the following prices for a jar of applesauce.

> 16-ounce jar: $1.19
> 24-ounce jar: $1.29
> 48-ounce jar: $2.69
> 64-ounce jar: $3.49

Which size is the best buy? That is, which size has the lowest unit price?

_____

Objective 1 Practice Exercises

For extra help, see Examples 1–2 on pages 174–175 of your text.

Write a ratio for each word phrase. Write fractions in lowest terms.

1. 8 men to 3 men

1. _____

2. 9 dollars to 48 quarters

2. _____

A supermarket was surveyed and the following prices were charged for items in various sizes. Find the best buy (based on price per unit) for each of the following items.

3.   Trash bags   3. _____
      10-count box: $2.89
      20-count box: $5.29
      45-count box: $6.69
      85-count box: $13.99

Objective 2   Solve proportions.

Video Examples

Review this example for Objective 2:

4.   Solve the equation $\dfrac{n-1}{3} = \dfrac{2n+1}{4}$.

$$\frac{n-1}{3} = \frac{2n+1}{4}$$

$$3(2n+1) = 4(n-1)$$

$$6n+3 = 4n-4$$

$$2n+3 = -4$$

$$2n = -7$$

$$n = -\frac{7}{2}$$

A check confirms that the solution is $-\dfrac{7}{2}$, so the

solution set is $\left\{-\dfrac{7}{2}\right\}$.

Now Try:

4.   Solve the equation

$$\frac{2x+1}{2} = \frac{7x+3}{9}.$$

_____

Name: _____ Date: _____

Instructor: _____ Section: _____

Objective 2 Practice Exercises

For extra help, see Examples 3–4 on page 176 of your text.

Solve each equation.

4. $\dfrac{z}{20} = \dfrac{25}{125}$

    4. _____

5. $\dfrac{m}{5} = \dfrac{m-2}{2}$

    5. _____

6. $\dfrac{z+1}{4} = \dfrac{z+7}{2}$

    6. _____

Objective 3   Solve applied problems using proportions.

Video Examples

Review this example for Objective 3:

5. If four pounds of fertilizer will cover 50 square feet of garden, how many pounds would be needed for 125 square feet?

   To solve this problem, set up a proportion, with pounds in the numerator and square feet in the denominator.

Now Try:

5. Margie earns $168.48 in 26 hours. How much does she earn in 40 hours?

    _____

$$\frac{4}{50} = \frac{x}{125}$$

$$4(125) = 50x$$

$$500 = 50x$$

$$10 = x$$

10 lb of fertilizer are needed.

Objective 3 Practice Exercises

For extra help, see Example 5 on page 177 of your text.

Solve each problem.

7. On a road map, 6 inches represents 50 miles. How   7. _____
    many inches would represent 125 miles?

8. If 12 rolls of tape cost $4.60, how much will 15 rolls   8. _____
    cost?

9. A garden service charges $30 to install 50 square   9. _____
    feet of sod. Find the charge to install 225 square feet.

Objective 4   Find percents and percentages.

Video Examples

Review these examples for Objective 4:

6.   Solve each problem.

    a. What is 18% of 700?

    Let n = the number. The word of indicates multiplication.

$$
\begin{array}{ccccc}
\text{What} & \text{is} & 18\% & \text{of} & 700 \\
\downarrow & \downarrow & \downarrow & \downarrow & \downarrow \\
n & = & 0.18 & \cdot & 700
\end{array}
$$
$$n = 126$$

    Thus, 126 is 18% of 700.

    b. 54% of what number is 162?

$$
\begin{array}{cccccc}
54\% & \text{of} & \text{what number} & \text{is} & 162 \\
\downarrow & \downarrow & & & \downarrow & \downarrow \\
0.54 & \cdot & n & & = & 162
\end{array}
$$
$$n = \frac{162}{0.54}$$
$$n = 300$$

    54% of 300 is 162.

Now Try:

6.   Solve each problem.

    a. What is 35% of 400?

    _____

    b. 42% of what number is 399?

    _____

Objective 4 Practice Exercises

For extra help, see Examples 6–7 on page 178 of your text.

Answer each question about percent.

10.   What is 2.5% of 3500?   10. _____

11.   What percent of 5200 is 104?   11. _____

Solve the problem.

12.   Paul recently bought a duplex for $144,000. He expects to earn $6120 per year on this investment. What percent of the purchase price will he earn?   12. _____

Chapter 2 EQUATIONS, INEQUALITIES, AND APPLICATIONS

2.7   Solving Linear Inequalities

Learning Objectives
1   Graph intervals on a number line.
2   Use the addition property of inequality.
3   Use the multiplication property of inequality.
4   Solve linear inequalities.
5   Solve applied problems using inequalities.
6   Solve linear inequalities with three parts.

Key Terms

Use the vocabulary terms listed below to complete each statement in exercises 1–5.

     inequalities   **interval**   **interval notation**

     linear inequality   **three-part inequality**

1.   An inequality that says that one number is between two other numbers is a(n)_____.

2.   A portion of a number line is called a(n) _____.

3.   A(n) _____ can be written in the form $Ax + B < C$, $Ax + B \leq C$, $Ax + B > C$, or $Ax + B \geq C$, where A, B, and C are real numbers with $A \neq 0$.

4.   Algebraic expressions related by $<, \leq, >,$ or $\geq$ are called _____.

5.   The _____ for $a \leq x < b$ is $[a, b)$.

Objective 1   Graph intervals on a number line.

Video Examples

Review this example for Objective 1:
1.   Graph $x > -3$.

    The statement $x > -3$ says that x can represent any value greater than –3, but cannot equal –3, written $(-3, \infty)$. We graph this interval by placing a parenthesis at –3 and drawing an arrow to the right. The parenthesis indicates that –3 is not part of the graph.

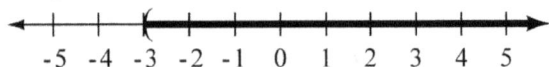

Now Try:
1.   Graph $x > -1$.

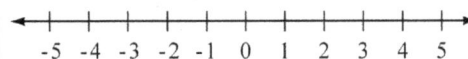

Name: Date:
Instructor: Section:

Objective 1 Practice Exercises

For extra help, see Examples 1–2 on page 187 of your text.

Write each inequality in interval notation and graph the interval.

1. $3 < a$

1. _____
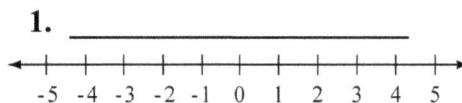

2. $y \geq -2$

2. _____
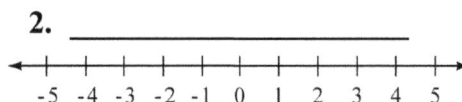

3. $x < -4$

3. _____
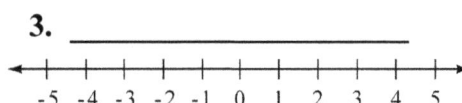

Objective 2   Use the addition property of inequality.

Video Examples

Review this example for Objective 2:

3.   Solve $8 + 4k \geq 3k + 3$ and graph the solution set.

$$8 + 4k \geq 3k + 3$$
$$8 + 4k - 3k \geq 3k + 3 - 3k$$
$$8 + k \geq 3$$
$$8 + k - 8 \geq 3 - 8$$
$$k \geq -5$$

The solution set, $[-5, \infty)$ is graphed below.

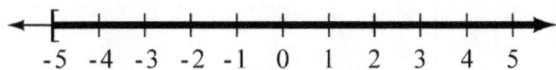

Now Try:

3.   Solve $5 + 9k \geq 8k + 2$ and graph the solution set.

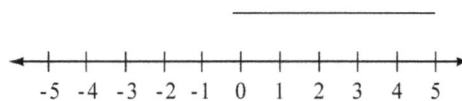

Objective 2 Practice Exercises

For extra help, see Example 3 on page 188 of your text.

Solve each inequality. Write the solution set in interval notation and then graph it.

4. $5a + 3 \leq 6a$

4. _____
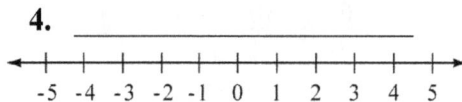

5. $6 + 3x < 4x + 4$

5. _____

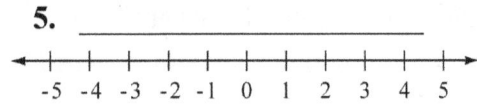

6. $3 + 5p \leq 4p + 3$

6. _____

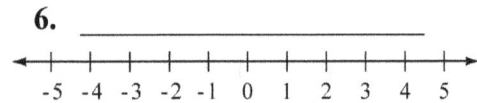

Objective 3   Use the multiplication property of inequality.

Video Examples

Review these examples for Objective 3:

4. Solve each inequality, and graph the solution set.

a. $6x < -24$

We divide each side by 6.
$$6x < -24$$
$$\frac{6x}{6} < \frac{-24}{6}$$
$$x < -4$$

The graph of the solution set $(-\infty, -4)$, is shown below.

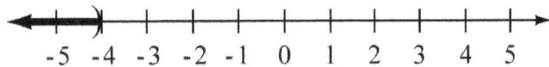

b. $-6x \geq 30$

Here each side of the inequality must be divided by –6, a negative number, which does require changing the direction of the inequality symbol.
$$-6x \geq 30$$
$$\frac{-6x}{-6} \leq \frac{30}{-6}$$
$$x \leq -5$$

The solution set, $(-\infty, -5]$, is graphed below.

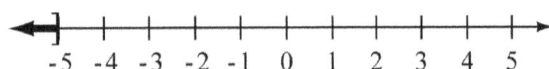

Now Try:

4. Solve each inequality, and graph the solution set.

a. $8x \leq -40$

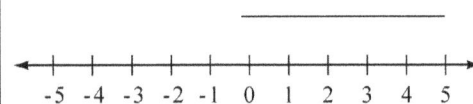

b. $-9t > 36$

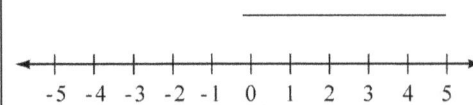

Name:
Instructor:

Date:
Section:

Objective 3 Practice Exercises

For extra help, see Example 4 on page 190 of your text.

Solve each inequality. Write the solution set in interval notation and then graph it.

7. $-2s < 4$

7. _____

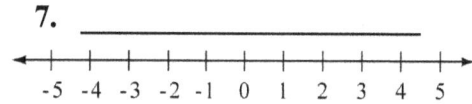

8. $4k \geq -16$

8. _____

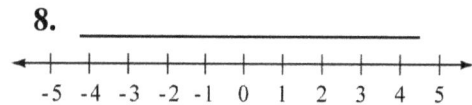

9. $-9m \geq -36$

9. _____

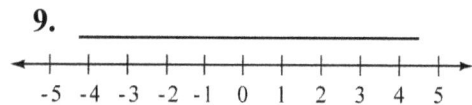

Objective 4   Solve linear inequalities.

Video Examples

Review these examples for Objective 4:

5. Solve $4x + 3 - 7 > -2x + 8 + 3x$. Graph the solution set.

Step 1 Combine like terms and simplify.
$$4x + 3 - 7 > -2x + 8 + 3x$$
$$4x - 4 > x + 8$$

Step 2 Use the addition property of inequality.
$$4x - 4 - x > x + 8 - x$$
$$3x - 4 > 8$$
$$3x - 4 + 4 > 8 + 4$$
$$3x > 12$$

Step 3 Use the multiplication property of inequality.
$$\frac{3x}{3} > \frac{12}{3}$$
$$x > 4$$

The solution set is $(4, \infty)$. The graph is shown below.

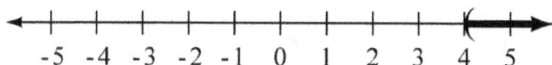

Now Try:

5. Solve $8x - 5 + 4 \geq 6x - 3x + 9$. Graph the solution set.

_____

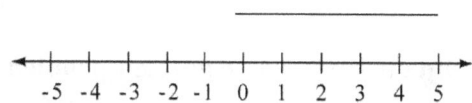

Objective 4 Practice Exercises

For extra help, see Example 5–7 on pages 191–192 of your text.

Solve each inequality. Write the solution set in interval notation and then graph it.

10. $4(y-3)+2>3(y-2)$

10. _____

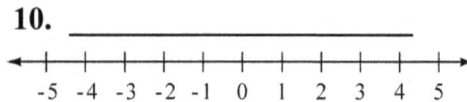
-5 -4 -3 -2 -1 0 1 2 3 4 5

11. $-3(m+2)+3\leq-4(m-2)-6$

11. _____

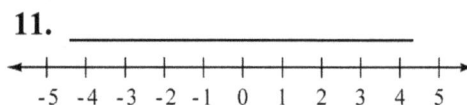
-5 -4 -3 -2 -1 0 1 2 3 4 5

12. $7(2-x)\leq-2(x-3)-x$

12. _____

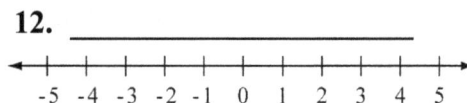
-5 -4 -3 -2 -1 0 1 2 3 4 5

Objective 5   Solve applied problems using inequalities.

Video Examples

Review this example for Objective 5:

8. Ruth tutors mathematics in the evenings in an office for which she pays $600 per month rent. If rent is her only expense and she charges each student $40 per month, how many students must she teach to make a profit of at least $1600 per month?

Step 1 Read the problem again.

Step 2 Assign a variable.
    Let x = the number of students.

Step 3 Write an inequality.
    $40x-600\geq1600$

Now Try:

8. Two sides of a triangle are equal in length, with the third side 8 feet longer than one of the equal sides. The perimeter of the triangle cannot be more than 38 feet. Find the largest possible value for the length of the equal sides.

_____

Copyright © 2018 Pearson Education, Inc.

Step 4 Solve.

$$40x - 600 + 600 \geq 1600 + 600$$

$$40x \geq 2200$$

$$\frac{40x}{40} \geq \frac{2200}{40}$$

$$x \geq 55$$

Step 5 State the answer. Ruth must have 55 or more students to have at least $1600 profit.

Step 6 Check. $40(55) - 600 = 1600$ Also, any number greater than 55 makes the profit greater than $1600.

Objective 5 Practice Exercises

For extra help, see Example 8 on page 193 of your text.

Solve each problem.

13.   Lauren has grades of 98 and 86 on her first two chemistry quizzes. What must she score on her third quiz to have an average of at least 91 on the three quizzes?

13. _____

14.   Nina has a budget of $230 for gifts for this year. So far she has bought gifts costing $47.52, $38.98, and $26.98. If she has three more gifts to buy, find the average amount she can spend on each gift and still stay within her budget.

14. _____

15.　If twice the sum of a number and 7 is subtracted　　**15.** _____
from three times the number, the result is more
than −9. Find all such numbers.

Objective 6　Solve linear inequalities with three parts.

Video Examples

Review these examples for Objective 6:

9.　Write the inequality $-4 \le x < 3$ in interval
notation, and graph the interval.

$-4 \le x < 3$
x is between −4 and 3 (excluding 3).
In interval notation, we write $[-4, 3)$.

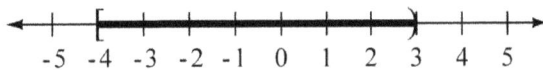

10a. Solve the inequality, and graph the solution set.

$3 \le 4x - 5 < 7$

$3 \le 4x - 5 < 7$
$3 + 5 \le 4x - 5 + 5 < 7 + 5$
$8 \le 4x < 12$
$\dfrac{8}{4} \le \dfrac{4x}{4} < \dfrac{12}{4}$
$2 \le x < 3$

The solution set is $[2, 3)$. The graph is shown
below.

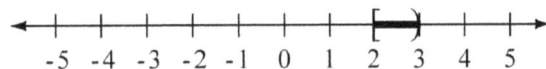

Now Try:

9.　Write the inequality
$-5 < x \le -1$ in interval
notation, and graph the interval.

_____

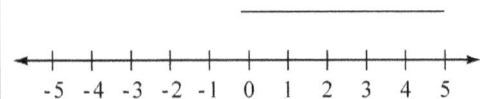

10a. Solve the inequality, and graph
the solution set.
$8 \le 6x - 4 < 20$

_____

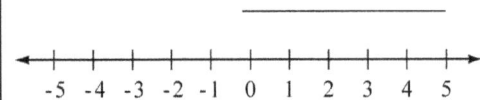

Name: _____ Date: _____

Instructor: _____ Section: _____

Objective 6 Practice Exercises

For extra help, see Examples 9–10 on pages 193–194 of your text.

Solve each inequality. Write the solution set in interval notation and then graph it.

16. $7 < 2x + 3 \leq 13$

16. _____

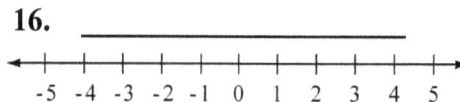

17. $-17 \leq 3x - 2 < -11$

17. _____

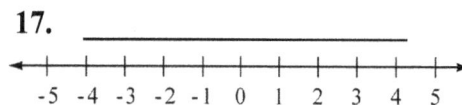

18. $1 < 3z + 4 < 19$

18. _____

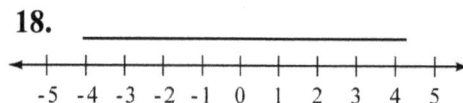

Chapter 3 GRAPHS OF LINEAR EQUATIONS AND INEQUALITIES IN TWO VARIABLES

3.1   Linear Equations and Rectangular Coordinates

Learning Objectives	
1	Interpret line graphs.
2	Write a solution as an ordered pair.
3	Decide whether a given ordered pair is a solution of a given equation.
4	Complete ordered pairs for a given equation.
5	Complete a table of values.
6	Plot ordered pairs.

Key Terms

Use the vocabulary terms listed below to complete each statement in exercises 1–13.

> line graph   linear equation in two variables
>
> ordered pair   table of values x-axis
>
> y-axis   rectangular (Cartesian) coordinate system
>
> origin   quadrants   plane   coordinates
>
> plot   scatter diagram

1. A _____ uses dots connected by lines to show trends.

2. An equation that can be written in the form $Ax + By = C$, where A, B, and C are real numbers and A, $B \neq 0$, is called a _____.

3. _____ are the numbers in the ordered pair that specify the location of a point on a rectangular coordinate system.

4. In a coordinate system, the horizontal axis is called the _____.

5. In a coordinate system, the vertical axis is called the _____.

6. A pair of numbers written between parentheses in which order is important is called a(n) _____.

7. Together, the x-axis and the y-axis form a _____.

8. A coordinate system divides the plane into four regions called _____.

9. The axis lines in a coordinate system intersect at the _____.

10. To _____ an ordered pair is to find the corresponding point on a coordinate system.

11. A graph of ordered pairs is called a _____.

12. A table showing selected ordered pairs of numbers that satisfy an equation is called a _____.

13. A flat surface determined by two intersecting lines is a _____.

Objective 1   Interpret line graphs.

Video Examples

The line graph shows the number of degrees awarded by a university for the years 2010–2015.

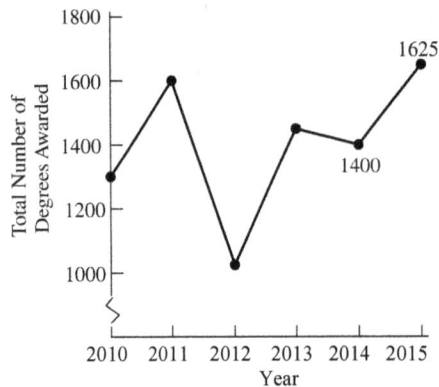

Review these examples for Objective 1:

1.

a. Between which years did the number of degrees awarded decrease?

The line between 2011 and 2012 and between 2013 and 2014 falls, so the number of degrees awarded decreased between 2011-2012 and 2013-2014.

b. Between which two years did the total number of degrees awarded show the smallest change?

The line between 2013 and 2014 falls the least, so the total number of degrees awarded declined the least from 2013 to 2014.

c. Estimate the total number of degrees awarded in 2012 and in 2015. About how many more degrees were awarded in 2015?

Move up from 2012 on the horizontal scale to the point plotted for 2012. This point is about 1000. So about 1000 degrees were awarded in 2012.

Now Try:

1.

a. Between which years did the number of degrees awarded increase?

_____

b. Between which two years did the total number of degrees awarded show the greatest decline?

_____

c. Estimate the total number of degrees awarded in 2010 and in 2011. About how many more degrees were awarded in 2011?

_____

Similarly, locate the point plotted for 2015. Moving across to the vertical scale, the graph indicates that the number of degrees awarded in 2015 was 1625.

Between 2012 and 2015, the increase was
$$1625 - 1000 = 625.$$

Objective 1 Practice Exercises

For extra help, see Example 1 on page 214 of your text.

The line graph shows the number of degrees awarded by a university for the years 2010–2015. Use this graph to answer exercises 1–3.

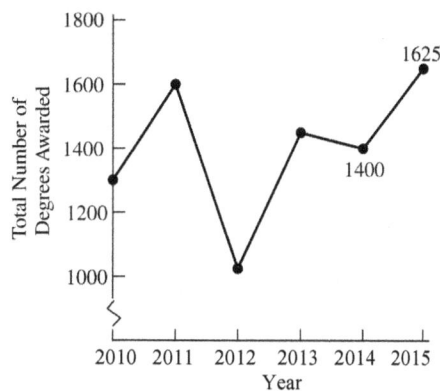

1.   Between what years did the number of degrees awarded have the least change?

1. _____

2.   Over which years was the number of degrees awarded more than 1400?

2. _____

3.   Estimate the average number of degrees awarded in 2011 and 2012. About how much did the number of degrees awarded decrease between 2011 and 2012?

3. _____

Objective 2   **Write a solution as an ordered pair.**

Objective 2 Practice Exercises

For extra help, see page 216 of your text.

Write each solution as an ordered pair.

4. $x = 4$ and $y = 7$ 4. _____

5. $y = \frac{1}{3}$ and $x = 0$ 5. _____

6. $x = 0.2$ and $y = 0.3$ 6. _____

Objective 3   **Decide whether a given ordered pair is a solution of a given equation.**

Video Examples

Review these examples for Objective 3:

2. Decide whether each ordered pair is a solution of the equation $4x + 5y = 40$.

   a. $(5,\ 4)$

Substitute 5 for *x* and 4 for *y* in the given equation.

$$4x + 5y = 40$$
$$4(5) + 5(4) \overset{?}{=} 40$$
$$20 + 20 \overset{?}{=} 40$$
$$40 = 40 \quad \text{True}$$

This result is true, so $(5,\ 4)$ is a solution of $4x + 5y = 40$.

   b. $(-3,\ 6)$

Substitute –3 for *x* and 6 for *y* in the given equation.

$$4x + 5y = 40$$
$$4(-3) + 5(6) \overset{?}{=} 40$$
$$-12 + 30 \overset{?}{=} 40$$
$$18 = 40 \quad \text{False}$$

This result is false, so $(-3,\ 6)$ is not a solution of $4x + 5y = 40$.

Now Try:

2. Decide whether each ordered pair is a solution of the equation $3x - 4y = 12$.

   a. $(8,\ 3)$

                 _____

   b. $(5, -4)$

                 _____

Objective 3 Practice Exercises

For extra help, see Example 2 on page 216 of your text.

Decide whether the given ordered pair is a solution of the given equation.

7. $4x - 3y = 10$; $(1, 2)$ 7. _____

8. $2x - 3y = 1$; $\left(0, \frac{1}{3}\right)$ 8. _____

9. $x = -7$; $(-7, 9)$ 9. _____

Objective 4   Complete ordered pairs for a given equation.

Video Examples

Review this example for Objective 4:
3a. Complete each ordered pair for the equation
   $y = 5x + 8$.

   $(3, \underline{\quad})$

   Replace x with 3.
   $y = 5x + 8$
   $y = 5(3) + 8$
   $y = 15 + 8$
   $y = 23$
   The ordered pair is $(3, 23)$.

Now Try:
3a. Complete each ordered pair for
   the equation $y = 4x - 7$.

   $(5, \underline{\quad})$

   _____

Objective 4 Practice Exercises

For extra help, see Example 3 on page 216 of your text.

For each of the given equations, complete the ordered pairs beneath it.

10. $y = 2x - 5$

     (a) $(2, \ )$

     (b) $(0, \ )$

     (c) $(\ , 3)$

     (d) $(\ , -7)$

     (e) $(\ , 9)$

10.

(a) _____

(b) _____

(c) _____

(d) _____

(e) _____

11. $y = 3 + 2x$

     (a) $(-4, \ )$

     (b) $(2, \ )$

     (c) $(\ , 0)$

     (d) $(-2, \ )$

     (e) $(\ , -7)$

11.

(a) _____

(b) _____

(c) _____

(d) _____

(e) _____

Name: Date:

Instructor: Section:

Objective 5   Complete a table of values.

Video Examples

Review this example for Objective 5:

4a. Complete the table of values for the equation. Then write the results as ordered pairs.

$2x - 3y = 6$

x	y
9	
6	
	-2
	8

From the table, we can write the ordered pairs:
(9, ____), (6, ____), (____, –2), (____, 8).
From the first row of the table, let $x = 9$ in the equation. From the second row of the table, let $x = 6$.

If $x = 9$,	If $x = 6$,
$2x - 3y = 6$	$2x - 3y = 6$
$2(9) - 3y = 6$	$2(6) - 3y = 6$
$18 - 3y = 6$	$12 - 3y = 6$
$-3y = -12$	$-3y = -6$
$y = 4$	$y = 2$

The first two ordered pairs are (9, 4) and (6, 2).

From the third and fourth rows of the table, let $y = -2$ and $y = 8$, respectively.

If $y = -2$,	If $y = 8$,
$2x - 3y = 6$	$2x - 3y = 6$
$2x - 3(-2) = 6$	$2x - 3(8) = 6$
$2x + 6 = 6$	$2x - 24 = 6$
$2x = 0$	$2x = 30$
$x = 0$	$x = 15$

The last two ordered pairs are (0, –2) and (15, 8). The completed table and corresponding ordered pairs follow.

x	y	Ordered pairs
9	4	→ (9, 4)
6	2	→ (6, 2)
0	–2	→ (0, –2)
15	8	→ (15, 8)

Now Try:

4a. Complete the table of values for the equation. Then write the results as ordered pairs.

$4x - y = 8$

x	y
1	
5	
	0
	4

_____

Name: Date:

Instructor: Section:

Objective 5 Practice Exercises

For extra help, see Example 4 on page 217 of your text.

Complete each table of values. Write the results as ordered pairs.

12. $2x + 5 = 7$ 12. _____

x	y
	−3
	0
	5

13. $y - 4 = 0$ 13. _____

x	y
−4	
0	
6	

14. $4x + 3y = 12$ 14. _____

x	y
0	
	0
	−1

Objective 6   Plot ordered pairs.

Video Examples

Review these examples for Objective 6:

5.   Plot the given points in a coordinate system.

(5, 4)   (−2,−1)   (−3, 5)   (4,−2)
(1,−2.5)   (3, 0)   (0, 4)

Step 1 Move right or left the number of units that correspond to the *x*-coordinate in the ordered pair—right if the *x*-coordinate is positive and left if it is negative.

Step 2 Then turn and move up or down the number of units that corresponds to the *y*-coordinate—up if the *y*-coordinate is positive or down if it is negative.

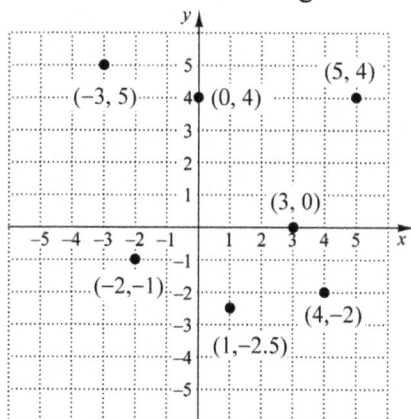

Now Try:

5.   Plot the given points in a coordinate system.

(2, 4)   (−5, 1)   (−4,−2)
(3,−5)   (2,−1.5)   (6, 0)
(0,−6)

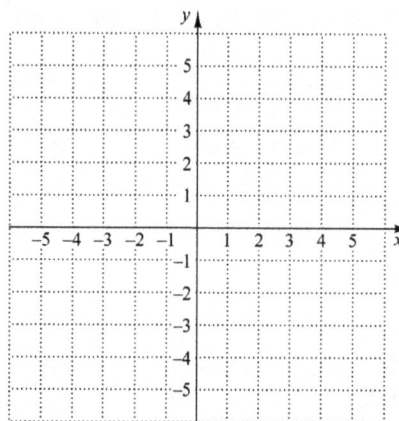

An accountant keeps track of the balance for a business loan given by the linear equation
$$y = 137.24x + 9879,$$
where *x* is the number of months, and *y* is the balance due. For this loan, payment is expected in full at the end of the term.

6.

   a. Complete the table of values for the given linear equation.

x (months)	y (balance due)
1	
3	
10	

To find *y* for *x* = 1, substitute into the equation.

6.

   a. Complete the table of values for the given linear equation.

x (months)	y (balance due)
2	
5	
11	

———————

$$y = 137.24x + 9879$$

$$y = 137.24(1) + 9879$$

$$y = 10,016.24$$

x (months)	y (balance due)
1	10,016.24
3	10,290.72
10	11,251.40

So the ordered pairs are (1, 10,016.24), (3, 10,290.72), and (10, 11,251.40).

b. Graph the ordered pairs found in part (a).

A graph of ordered pairs of data is a scatter diagram.

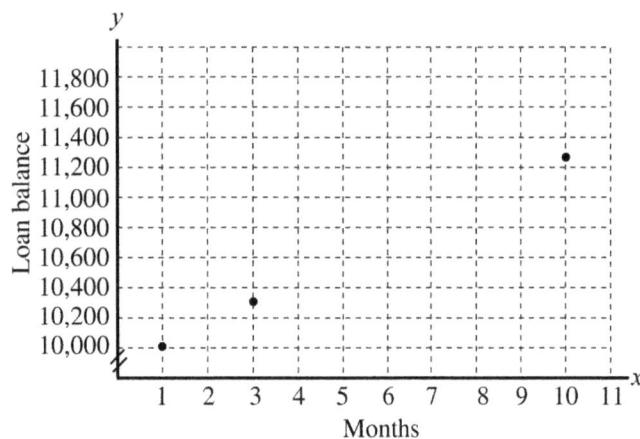

Months

b. Graph the ordered pairs found in part (a).

Objective 6 Practice Exercises

For extra help, see Examples 5–6 on pages 219–220 of your text.

Plot the each ordered pair on a coordinate system.

15. $(0, -2)$

15. _____

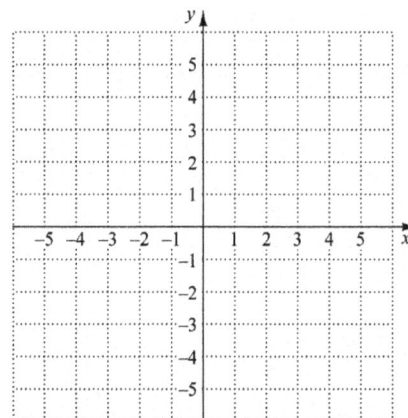

16. $(-3,\ 4)$ **16.** _____

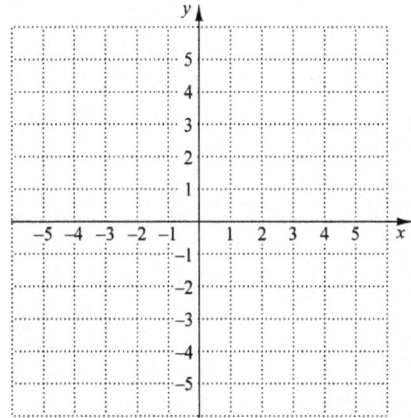

17. $(2, -5)$ **17.** _____

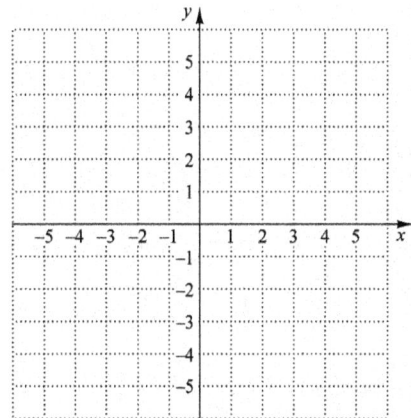

Chapter 3 GRAPHS OF LINEAR EQUATIONS AND INEQUALITIES IN TWO VARIABLES

3.2   Graphing Linear Equations in Two Variables

Learning Objectives	
1	Graph linear equations by plotting ordered pairs.
2	Find intercepts.
3	Graph linear equations of the form $Ax + By = 0$.
4	Graph linear equations of the form $y = b$ or $x = a$.
5	Use a linear equation to model data.

Key Terms

Use the vocabulary terms listed below to complete each statement in exercises 1–4.

graph   graphing   *y*-intercept   *x*-intercept

1. If a graph intersects the *y*-axis at *k*, then the _____ is (0, *k*).

2. If a graph intersects the *x*-axis at *k*, then the _____ is (*k*, 0).

3. The process of plotting the ordered pairs that satisfy a linear equation and drawing a line through them is called _____.

4. The set of all points that correspond to the ordered pairs that satisfy the equation is called the _____ of the equation.

Objective 1   Graph linear equations by plotting ordered pairs.

Video Examples

Review this example for Objective 1:

2.   Graph $2x + 3y = 6$.

First let $x = 0$ and then let $y = 0$ to determine two ordered pairs.

$$2(0) + 3y = 6 \quad \bigg| \quad 2x + 3(0) = 6$$
$$0 + 3y = 6 \quad \bigg| \quad 2x + 0 = 6$$
$$3y = 6 \quad \bigg| \quad 2x = 6$$
$$y = 2 \quad \bigg| \quad x = 3$$

The ordered pairs are (0, 2) and (3, 0). Find a third ordered pair by choosing a number other than 0 for *x* or *y*. We choose $y = 4$.

Now Try:

2.   Graph $x + y = 3$.

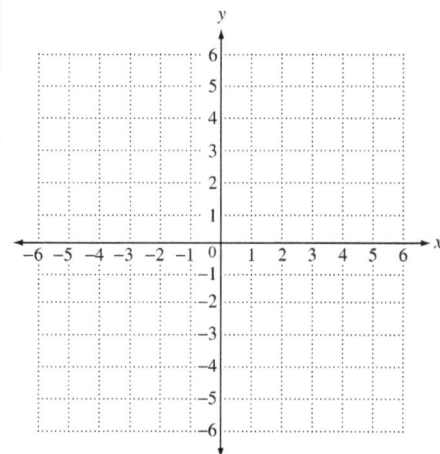

Copyright © 2018 Pearson Education, Inc.

$$2x + 3(4) = 6$$

$$2x + 12 = 6$$

$$2x = -6$$

$$x = -3$$

This gives the ordered pair (–3, 4). We plot the three ordered pairs (0, 2), (3, 0), and (–3, 4) and draw a line through them.

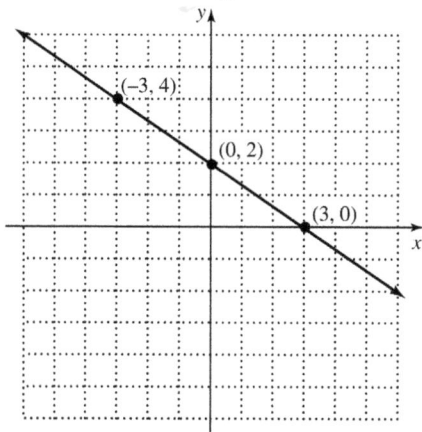

Objective 1 Practice Exercises

For extra help, see Examples 1–2 on pages 229–230 of your text.

Complete the ordered pairs for each equation. Then graph the equation by plotting the points and drawing a line through them.

1. $y = 3x - 2$

    (0,   )

    (, 0)

    (2,   )

1.

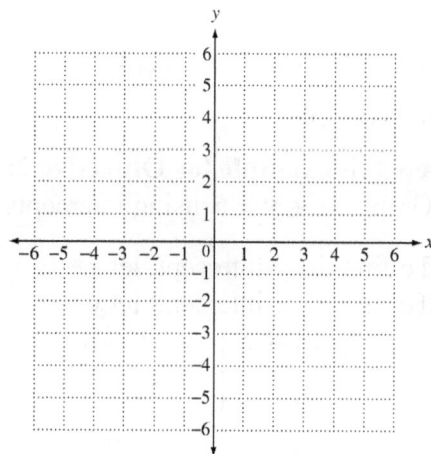

2. $x - y = 4$

$(0, \quad)$

$(\quad, 0)$

$(-2, \quad)$

2.

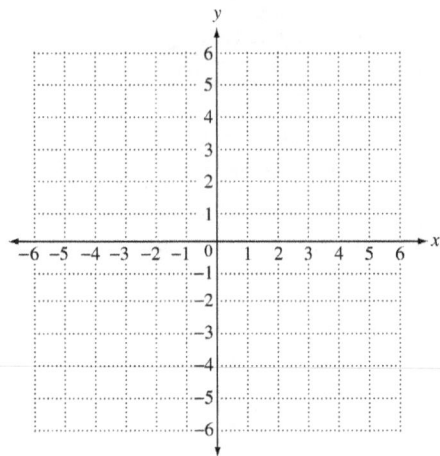

3. $x = 2y + 1$

$(0, \quad)$

$(\quad, 0)$

$(\quad, -2)$

3.

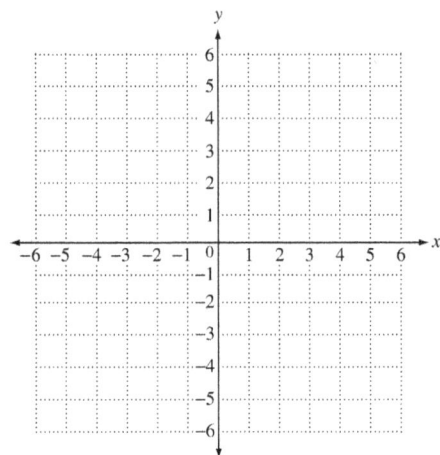

Objective 2   Find intercepts.

Video Examples

Review this example for Objective 2:

3. Graph $3x + y = 6$ using intercepts.

To find the y-intercept, let $x = 0$.
To find the x-intercept, let $y = 0$.

$$3(0) + y = 6 \quad \bigm| \quad 3x + 0 = 6$$
$$0 + y = 6 \quad \bigm| \quad 3x = 6$$
$$y = 6 \quad \bigm| \quad x = 2$$

The intercepts are $(0, 6)$ and $(2, 0)$. To find a third point, as a check, we let $x = 1$.

$$3(1) + y = 6$$
$$3 + y = 6$$
$$y = 3$$

Now Try:

3. Graph $5x - 2y = -10$ using intercepts.

This gives the ordered pair (1, 3).

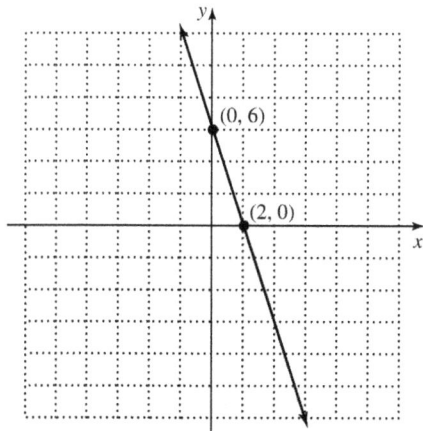

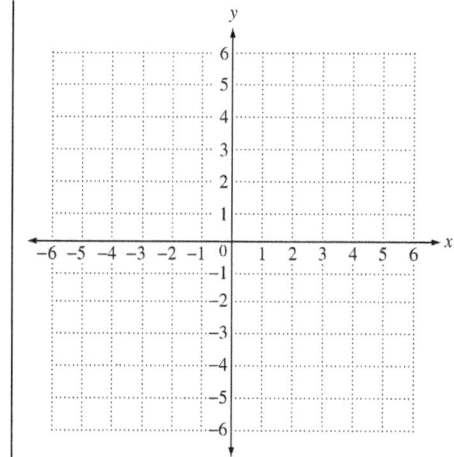

Objective 2 Practice Exercises

For extra help, see Examples 3–4 on pages 230–231 of your text.

Find the intercepts for each equation. Then graph the equation.

4. $y = \dfrac{2}{3}x - 2$

4.

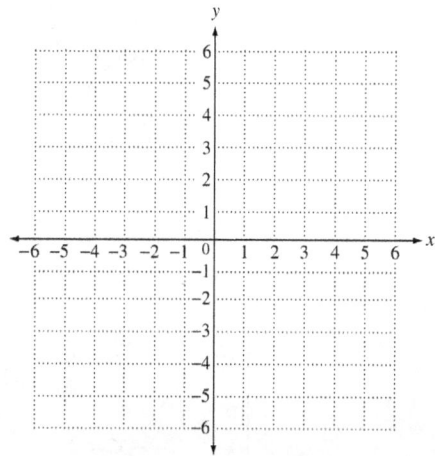

5. $4x - 7y = -8$

5.

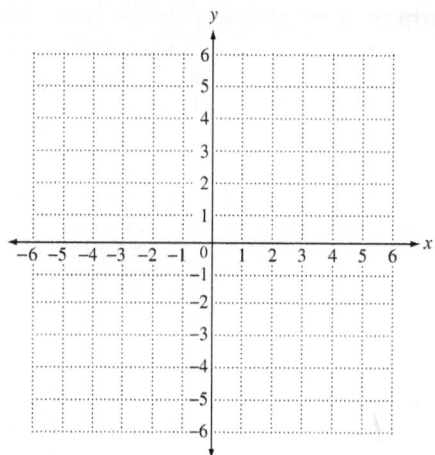

6. $y = 4x - 4$

6.

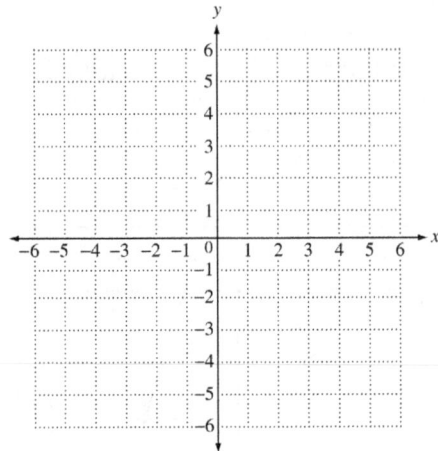

Objective 3   Graph linear equations of the form $Ax + By = 0$.

Video Examples

Review this example for Objective 3:

5.   Graph $x + 5y = 0$.

To find the y-intercept, let $x = 0$.
To find the x-intercept, let $y = 0$.

$$0 + 5y = 0 \quad | \quad x + 5(0) = 0$$
$$5y = 0 \quad | \quad x + 0 = 0$$
$$y = 0 \quad | \quad x = 0$$

The x- and y-intercepts are the same point $(0, 0)$. We must select two other values for x or y to find two other points. We choose $y = 1$ and $y = -1$.

$$x + 5(1) = 0 \quad | \quad x + 5(-1) = 0$$
$$x + 5 = 0 \quad | \quad x - 5 = 0$$
$$x = -5 \quad | \quad x = 5$$

We use $(-5, 1)$, $(0, 0)$, and $(5, -1)$ to draw the graph.

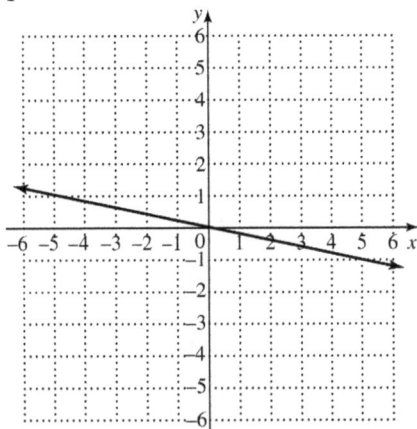

Now Try:

5.   Graph $3x - y = 0$.

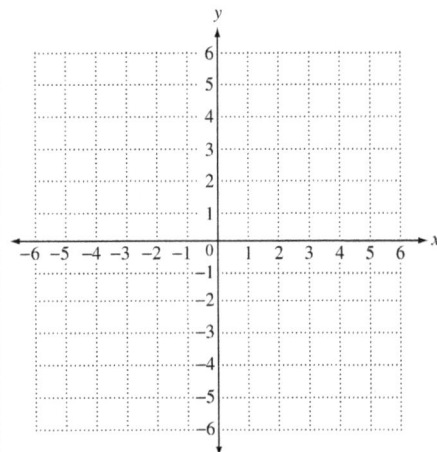

Name: Date:

Instructor: Section:

Objective 3 Practice Exercises

For extra help, see Example 5 on page 232 of your text.

Graph each equation.

7. $-3x - 2y = 0$ 7.

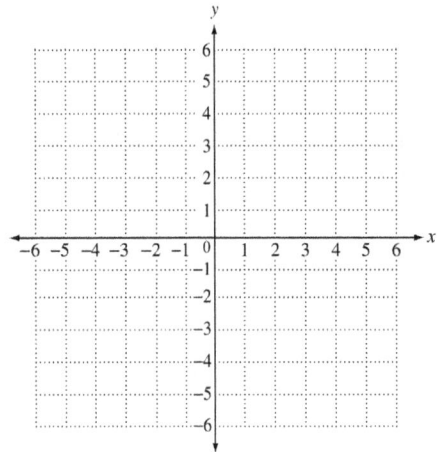

8. $x + y = 0$ 8.

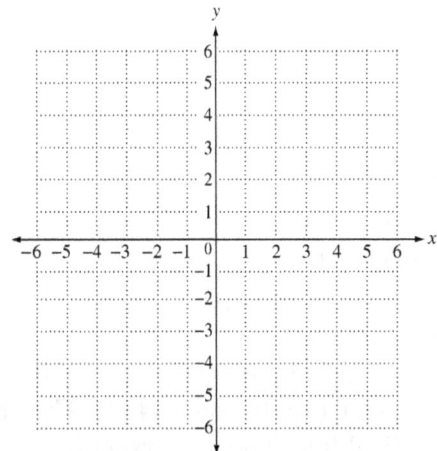

9. $y = 2x$ 9.

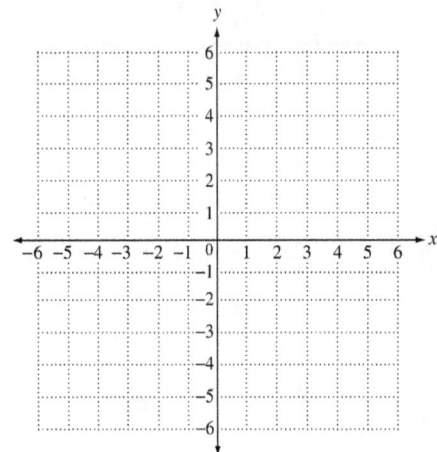

Objective 4   Graph linear equations of the form $y = b$ or $x = a$.

Video Examples

Review these examples for Objective 4:

6. Graph $y = -2$.

For any value of x, y is always -2. Three ordered pairs that satisfy the equation are $(-4, -2)$, $(0, -2)$ and $(2, -2)$. Drawing a line through these points gives the horizontal line. The y-intercept is $(0, -2)$. There is no x-intercept.

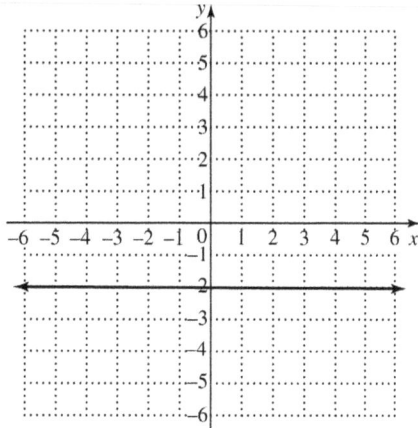

Now Try:

6. Graph $y = 4$.

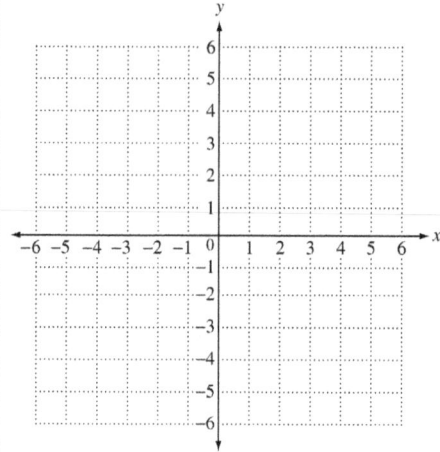

7. Graph $x + 4 = 0$.

First we subtract 4 from each side of the equation to get the equivalent equation $x = -4$. All ordered-pair solutions of this equation have x-coordinate -4.

Three ordered pairs that satisfy the equation are $(-4, -1)$, $(-4, 0)$, and $(-4, 3)$. The graph is a vertical line. The x-intercept is $(-4, 0)$. There is no y-intercept.

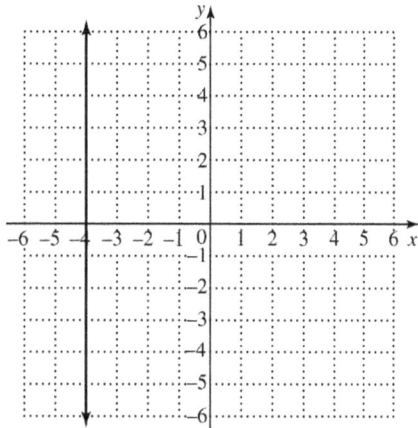

7. Graph $x = 0$.

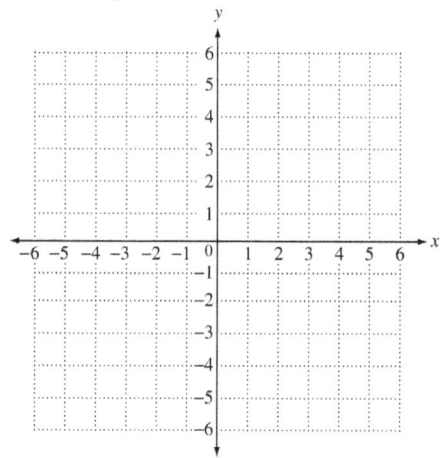

Name: Date:

Instructor: Section:

Objective 4 Practice Exercises

For extra help, see Examples 6–7 on page 233 of your text.

Graph each equation.

10. $x - 1 = 0$

10.

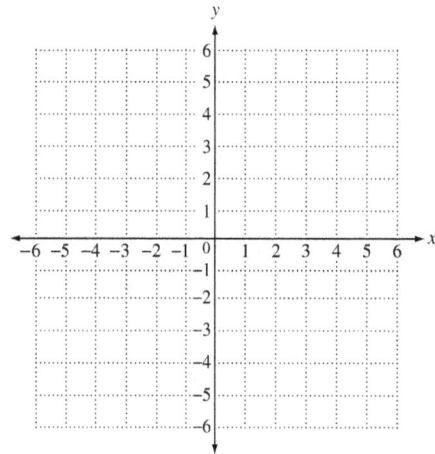

11. $y + 3 = 0$

11.

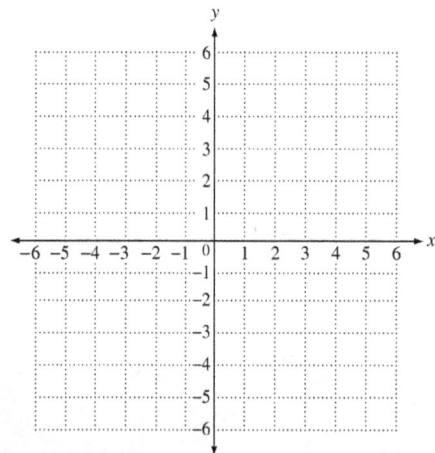

Objective 5   Use a linear equation to model data.

Video Examples

Review these examples for Objective 5:

8.   Every year sea turtles return to a certain group of islands to lay eggs. The number of turtle eggs that hatch can be approximated by the equation $y = -70x + 3260$, where y is the number of eggs that hatch and $x = 0$ representing 1990.

a. Use this equation to find the number of eggs that hatched in 1995, 2000, and 2005, and 2015.

Substitute the appropriate value for each year x to find the number of eggs hatched in that year.

For 1995:

$$y = -70(5) + 3260 \qquad 1995 - 1990 = 5$$

$$y = 2910 \text{ eggs} \qquad \text{Replace } x \text{ with 5.}$$

For 2000:

$$y = -70(10) + 3260 \qquad 2000 - 1990 = 10$$

$$y = 2560 \text{ eggs} \qquad \text{Replace } x \text{ with 10.}$$

For 2005:

$$y = -70(15) + 3260 \qquad 2005 - 1990 = 15$$

$$y = 2210 \text{ eggs} \qquad \text{Replace } x \text{ with 15.}$$

For 2015:

$$y = -70(25) + 3260 \qquad 2015 - 1990 = 25$$

$$y = 1510 \text{ eggs} \qquad \text{Replace } x \text{ with 25.}$$

b. Write the information from part (a) as four ordered pairs, and use them to graph the given linear equation.

Since x represents the year and y represents the number of eggs, the ordered pairs are $(5, 2910)$, $(10, 2560)$, $(15, 2210)$, and $(25, 1510)$.

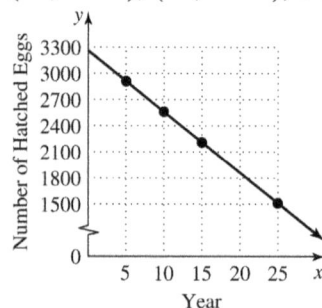

Now Try:

8.   Suppose that the demand and price for a certain model of calculator are related by the equation $y = 45 - \dfrac{3}{5}x$, where y is the price (in dollars) and x is the demand (in thousands of calculators).

a. Assuming that this model is valid for a demand up to 50,000 calculators, use this equation to find the price of calculators at each level of demand.

0 calculators _____

5000 calculators _____

20,000 calculators _____

45,000 calculators _____

b. Write the information from part (a) as four ordered pairs, and use them to graph the given linear equation.

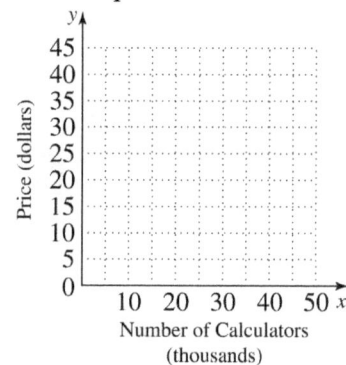

c. Use this graph and the equation to estimate the number of eggs that will hatch in 2010.

For 2010, $x = 20$. On the graph, find 20 on the horizontal axis, move up to the graphed line and then across to the vertical axis. It appears that in 2010, there were about 1900 eggs.

To use the equation, substitute 20 for x.

$$y = -70(20) + 3260$$

$$y = 1860 \text{ eggs}$$

This result for 2020 is close to our estimate of 1900 eggs from the graph.

c. Use this graph and the equation to estimate the price of 30,000 calculators.

_____

Objective 5 Practice Exercises

For extra help, see Example 8 on pages 234–235 of your text.

Solve each problem. Then graph the equation.

12. The profit y in millions of dollars earned by a small computer company can be approximated by the linear equation $y = 0.63x + 4.9$, where $x = 0$ corresponds to 2014, $x = 1$ corresponds to 2015, and so on. Use this equation to approximate the profit in each year from 2014 through 2017.

12. 2014 _____

    2015 _____

    2016 _____

    2017 _____

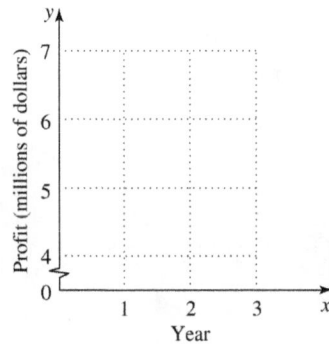

13. The number of band instruments sold by Elmer's Music Shop can be approximated by the equation $y = 325 + 42x$, where y is the number of instruments sold and x is the time in years, with $x = 0$ representing 2013. Use this equation to approximate the number of instruments sold in each year from 2013 through 2016.

13. 2013 _____

2014 _____

2015 _____

2016 _____

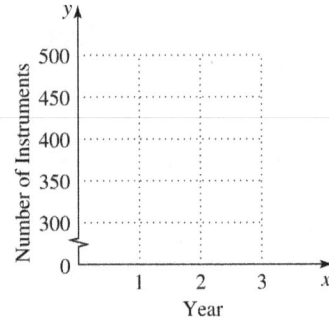

14. According to *The Old Farmer's Almanac*, the temperature in degrees Celsius can be determined by the equation $y = \frac{1}{3}x + 4$, where x is the number of cricket chirps in 25 seconds and y is the temperature in degrees Celsius. Use this equation to find the temperature when there are 48 chirps, 54 chirps, 60 chirps, and 66 chirps.

14. 48 _____

54 _____

60 _____

66 _____

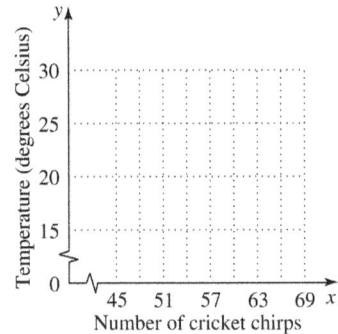

Chapter 3 GRAPHS OF LINEAR EQUATIONS AND INEQUALITIES IN TWO VARIABLES

3.3   The Slope of a Line

Learning Objectives
1   Find the slope of a line given two points.
2   Find the slope from the equation of a line.
3   Use slope to determine whether two lines are parallel, perpendicular, or neither.

Key Terms

Use the vocabulary terms listed below to complete each statement in exercises 1–5.

rise   **run**   **slope**   **parallel lines**

perpendicular lines

1.   Two lines that intersect in a 90° angle are called _____.

2.   The _____ of a line is the ratio of the change in y compared to the change in x when moving along the line from one point to another.

3.   The vertical change between two different points on a line is called the _____.

4.   Two lines in a plane that never intersect are called _____.

5.   The horizontal change between two different points on a line is called the _____.

Objective 1   Find the slope of a line given two points.

Video Examples

Review these examples for Objective 1:
1.   Find the slope of the line.

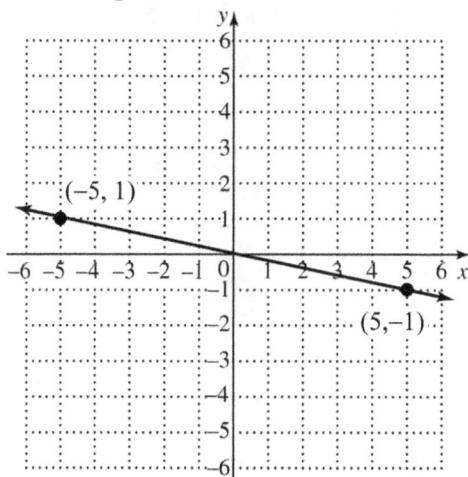

Now Try:
1.   Find the slope of the line.

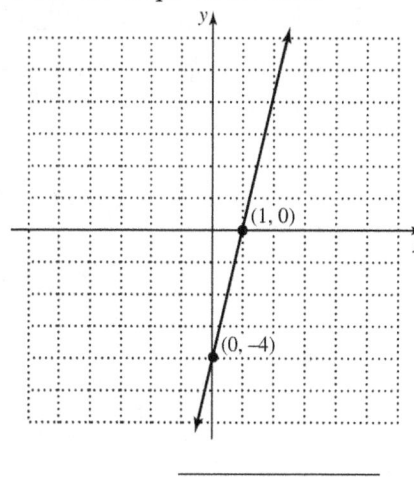

_____

We use the two points shown on the line. The vertical change is the difference in the y-values, or $-1 - 1 = -2$, and the horizontal change is the difference in the x-values or $5 - (-5) = 10$. Thus, the line has

$$\text{slope} = \frac{-2}{10}, \text{ or } -\frac{1}{5}.$$

3. Find the slope of the line passing through $(-9, 3)$ and $(4, 3)$.

$$(x_1, y_1) = (-9, 3) \text{ and } (x_2, y_2) = (4, 3)$$

$$m = \frac{y_2 - y_1}{x_2 - x_1} = \frac{3 - 3}{4 - (-9)} = \frac{0}{13} = 0$$

4. Find the slope of the line passing through $(-5, 3)$ and $(-5, 8)$.

$$(x_1, y_1) = (-5, 3) \text{ and } (x_2, y_2) = (-5, 8)$$

$$m = \frac{y_2 - y_1}{x_2 - x_1} = \frac{8 - 3}{-5 - (-5)} = \frac{5}{0} \text{ undefined slope}$$

2. Find the slope of each line.

a. The line passing through $(-5, 4)$ and $(2, -6)$

Apply the slope formula.
$$(x_1, y_1) = (-5, 4) \text{ and } (x_2, y_2) = (2, -6)$$

$$\text{slope } m = \frac{y_2 - y_1}{x_2 - x_1} = \frac{-6 - 4}{2 - (-5)}$$

$$= \frac{-10}{7}, \text{ or } -\frac{10}{7}$$

b. The line passing through $(-7, -3)$ and $(11, 8)$

Apply the slope formula.
$$(x_1, y_1) = (-7, -3) \text{ and } (x_2, y_2) = (11, 8)$$

$$\text{slope } m = \frac{y_2 - y_1}{x_2 - x_1} = \frac{8 - (-3)}{11 - (-7)} = \frac{11}{18}$$

3. Find the slope of the line passing through $(8, -5)$ and $(-7, -5)$.

_____

4. Find the slope of the line passing through $(9, 11)$ and $(9, -7)$.

_____

2. Find the slope of each line.

a. The line passing through $(-6, 7)$ and $(3, -9)$

_____

b. The line passing through $(-5, -2)$ and $(12, 7)$

_____

Objective 1 Practice Exercises

For extra help, see Examples 1–4 on pages 243–246 of your text.

Find the slope of the line through the given points.

1. (4, 3) and (3, 5)   1. _____

2. (−4, 6) and (−4, −1)   2. _____

3. (−3, 3) and (6, 3)   3. _____

Objective 2   Find the slope from the equation of a line.

Video Examples

Review these examples for Objective 2:	**Now Try:**
5. Find the slope of each line.	**5.** Find the slope of each line.
a. $4x - 3y = 7$	**a.** $7x - 4y = 8$

Step 1 Solve the equation for *y*.

$$4x - 3y = 7$$
$$-3y = -4x + 7$$
$$y = \frac{4}{3}x - \frac{7}{3}$$

Step 2 The slope is given by the coefficient of *x*,

so the slope is $\frac{4}{3}$.

_____

b. $6x + 3y = 1$

$3y = -6x + 1$

$y = -2x + \dfrac{1}{3}$

The slope of this line is given by the coefficient of x, which is -2.

c. $5y + x = 5$

$5y + x = 5$

$5y = -x + 5$

$y = -\dfrac{x}{5} + 1$

$y = -\dfrac{1}{5}x + 1$

The coefficient of x is $-\dfrac{1}{5}$, so the slope of the

line is $-\dfrac{1}{5}$.

b. $9x + 6y = 4$

_____

c. $8y + x = 16$

_____

Objective 2 Practice Exercises

For extra help, see Example 5 on page 247 of your text.

Find the slope of each line.

4. $7y - 4x = 11$ **4.** _____

5. $3y = 2x - 1$ **5.** _____

6. $y = -\dfrac{2}{5}x - 4$ **6.** _____

Objective 3   Use slope to determine whether two lines are parallel, perpendicular, or neither.

Video Examples

Review these examples for Objective 3:

6. Decide whether each pair of lines is parallel, perpendicular, or neither.

    b. $5x - y = 3$

            $15x - 3y = 12$

Solve each equation for y.

    $y = 5x - 3$

    $y = 5x - 4$

Both lines have slope 5, so the lines are parallel.

    a. $x + 3y = 8$ and $-3x + y = 5$

Find the slope of each line by first solving each equation for y.

$$3y = -x + 8 \qquad \big| \qquad y = 3x + 5$$

$$y = -\frac{1}{3}x + \frac{8}{3}$$

The slope is $-\frac{1}{3}$. $\big|$ The slope is 3.

Because the slopes are not equal, the lines are not parallel.

Check the product of the slopes: $-\frac{1}{3}(3) = -1$.

The two lines are perpendicular because the product of their slopes is -1.

Now Try:

6. Decide whether each pair of lines is parallel, perpendicular, or neither.

    b. $2x - 4y = 7$

          $3x - 6y = 8$

      _____

    a. $9x - y = 7$ and $x + 9y = 11$

      _____

Objective 3 Practice Exercises

For extra help, see Example 6 on page 249 of your text.

In each pair of equations, give the slope of each line, and then determine whether the two lines are **parallel,** **perpendicular,** *or* **neither.**

    7. $-x + y = -7$

          $x - y = -3$

    7. _____

8. $4x + 2y = 8$
       $x + 4y = -3$

8. _____

9. $9x + 3y = 2$
       $x - 3y = 5$

9. _____

Chapter 3 GRAPHS OF LINEAR EQUATIONS AND INEQUALITIES IN TWO VARIABLES

3.4   Slope-Intercept Form of a Linear Equation

Learning Objectives
1   Use slope-intercept form of the equation of a line.
2   Graph a line by using its slope and a point on the line.
3   Write an equation of a line by using its slope and any point on the line.
4   Graph and write equations of horizontal and vertical lines.

Key Terms

Use the vocabulary terms listed below to complete each statement in exercises 1−4.

   slope   ***y*-intercept**   **slope-intercept form**   **standard form**

1.   A linear equation in the form $y = mx + b$ is written in

      _____.

2.   A linear equation in the form $Ax + By = C$ is written in

      _____.

3.   In the linear equation $y = mx + b$, the variable b represents the

      _____ of the line.

4.   In the linear equation $y = mx + b$, the variable m represents the

      _____ of the line.

Objective 1   Use the slope-intercept form of the equation of a line.

Video Examples

Review these examples for Objective 1:

1.   Identify the slope and *y*-intercept of the line with each equation.

   a. $y = -8x + 7$

   The slope is –8, and the *y*-intercept is (0, 7).

   c. $y = 11x$

   The equation can be written as
      $y = 11x + 0$.
   The slope is 11, and the *y*-intercept is (0, 0).

Now Try:

1.   Identify the slope and *y*-intercept of the line with each equation.

   a. $y = -12x + 6$

      _____

   c. $y = 23x$

      _____

Objective 1 Practice Exercises

For extra help, see Example 1 on page 256 of your text.

Identify the slope and y-intercept of the line with each equation.

1. $y = \dfrac{3}{2}x - \dfrac{2}{3}$ 1. _____

                                                              _____

2. $y = -4x$ 2. _____

                                                              _____

Write an equation of the line with the given slope and y-intercept.

3.   slope -3; y-intercept $(0, 3)$ 3. _____

Objective 2   Graph a line using its slope and a point on the line.

Video Examples

Review these examples for Objective 2:

2b. Graph the equation by using the slope and y-intercept.

$$2x - 3y = 6$$

Step 1 Solve for y to write the equation in slope-intercept form.

$$2x - 3y = 6$$
$$-3y = -2x + 6$$
$$y = \frac{2}{3}x - 2$$

Step 2 The y-intercept is $(0, -2)$. Graph this point.

Step 3 The slope is $\dfrac{2}{3}$. By definition,

$$\text{slope } m = \frac{\text{change in } y \text{ (rise)}}{\text{change in } x \text{ (run)}} = \frac{2}{3}$$

From the y-intercept, count up 2 units and to the right 3 units to obtain the point $(3, 0)$.

Now Try:

2b. Graph the equation by using the slope and y-intercept.

$$y = \frac{2}{3}x$$

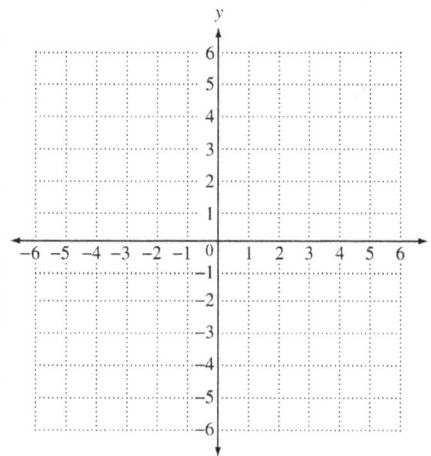

Step 4 Draw the line through the points (0,–2) and (3, 0) to obtain the graph.

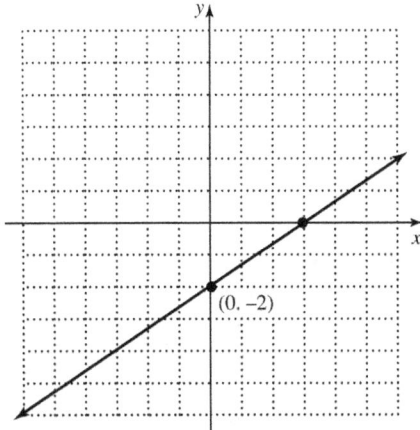

3. Graph the line passing through the point (1,–3), with slope $-\dfrac{5}{2}$.

First, locate the point (1,–3). Then write the slope $-\dfrac{5}{2}$ as

$$\text{slope } m = \frac{\text{change in } y \text{ (rise)}}{\text{change in } x \text{ (run)}} = \frac{5}{-2}.$$

Locate another point on the line by counting up 5 units from (1, –3), and then to the left 2 units. Finally, draw the line through this new point, (–1, 2).

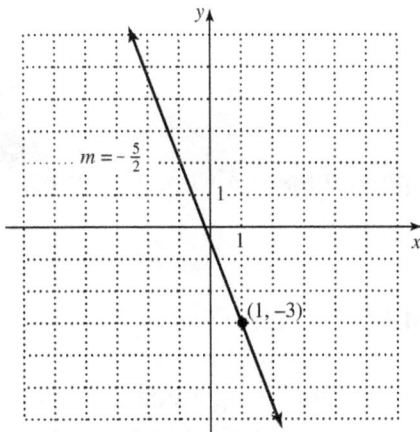

3. Graph the line passing through the point (2, 2), with slope $\dfrac{1}{3}$.

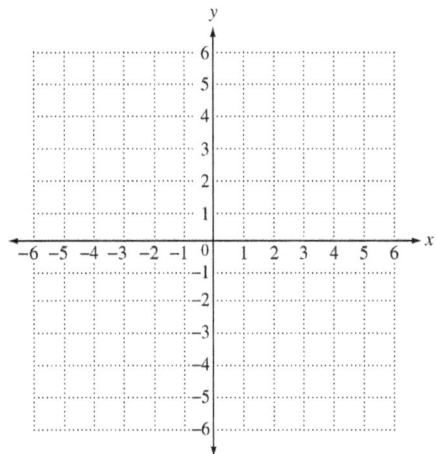

Name: _____   Date: _____

Instructor: _____   Section: _____

Objective 2 Practice Exercises

For extra help, see Examples 2–3 on pages 257–258 of your text.

Graph each equation by using the slope and y-intercept.

4. $4x - y = 4$ **4.**

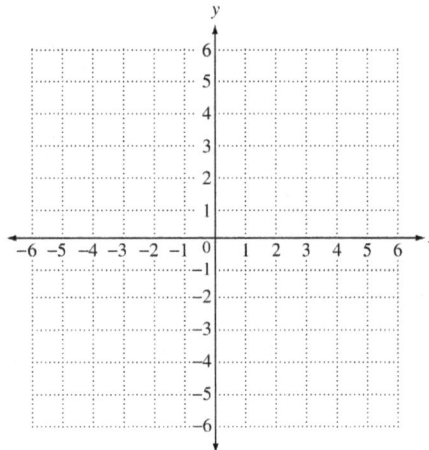

5. $y = -3x + 6$ **5.**

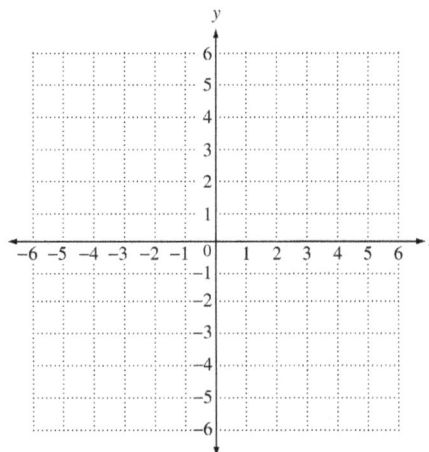

Objective 3   Write an equation of a line using its slope and any point on the line.

Video Examples

Review these examples for Objective 3:

4. Write an equation, in slope-intercept form of the line passing through the given point and having the given slope.

a. $(0, -3)$; $m = \dfrac{2}{5}$

Because the point $(0, -3)$ is the y-intercept, $b = -3$. We can substitute this value for b and the given

Now Try:

4. Write an equation, in slope-intercept form of the line passing through the given point and having the given slope.

a. $(0, -6)$; $m = \dfrac{3}{4}$

Copyright © 2018 Pearson Education, Inc.

slope $m = \dfrac{2}{5}$ directly into slope-intercept form

$y = mx + b$ to write an equation.

$$y = mx + b$$

$$y = \dfrac{2}{5}x + (-3)$$

$$y = \dfrac{2}{5}x - 3$$

b. $(2, 9),\ m = 5$

Since the line passes through the point $(2, 9)$, we can substitute $x = 2$, $y = 9$, and slope $m = 5$ into $y = mx + b$ and solve for b.

$$y = mx + b$$

$$9 = 5(2) + b$$

$$-1 = b$$

Now substitute the values of m and b into slope-intercept form.

$$y = mx + b$$

$$y = 5x - 1$$

b. $(-1, 4),\ m = 6$

Objective 3 Practice Exercises

For extra help, see Example 4 on pages 259 of your text.

Write an equation for the line passing through the given point and having the given slope. Write the equations in slope-intercept form, if possible.

6. $(-3,\ 4);\ m = -\dfrac{3}{5}$

7. _____

7. $(-4, -3);\ m = -2$

8. _____

8. $(0,\ 2);\ m = -\dfrac{3}{2}$

9. _____

Name: _____   Date: _____

Instructor: _____   Section: _____

Objective 4   Graph and write equations of horizontal and vertical lines.

Video Examples

Review this example for Objective 4:

5a. Graph the line passing through the given point and having the given slope.

  a. $(6, -2)$, $m = 0$

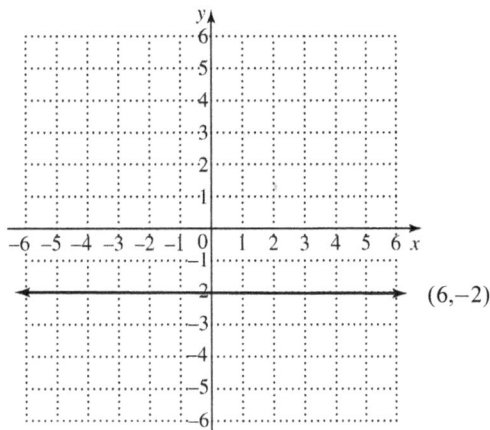

(6,−2)

Now Try:

5a. Graph the line passing through the given point and having the given slope.

  a. $(-4, 4)$, $m = 0$

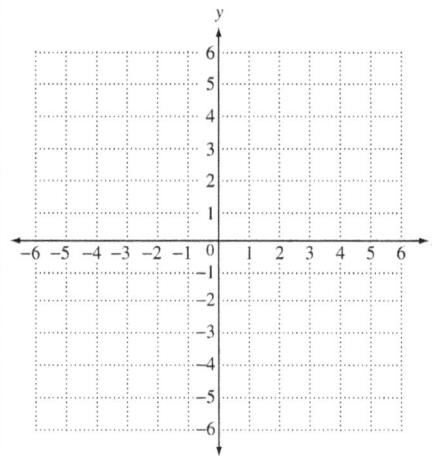

Objective 4 Practice Exercises

For extra help, see Examples 5–6 on pages 259–260 of your text.

Graph each line passing through the given point and having the given slope.

  9. $(1, 4)$, undefined slope

  9.

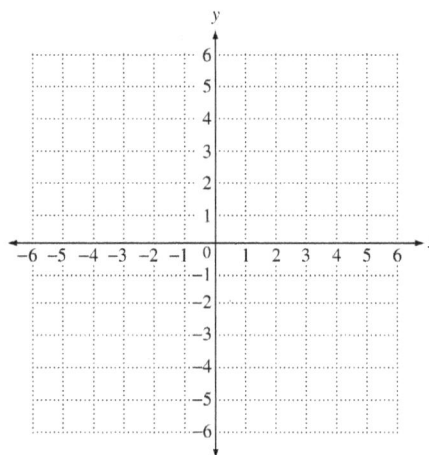

10. $(-2, -2)$; $m = 0$ **10.**

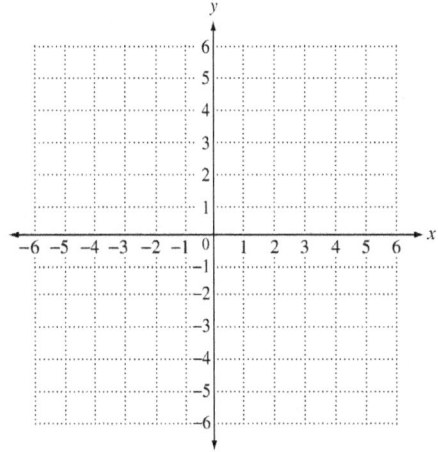

Write an equation of the line passing through the given point and having the given slope.

11. $(3, -5)$; undefined slope   **11.** _____

Chapter 3 GRAPHS OF LINEAR EQUATIONS AND INEQUALITIES IN TWO VARIABLES

3.5 Point-Slope Form of a Linear Equation and Modeling

Learning Objectives	
1	Use point-slope form to write an equation of a line.
2	Write an equation of a line using two points on the line.
3	Write an equation of a line that fits a data set.

Key Terms

Use the vocabulary terms listed below to complete each statement in exercises 1–3.

slope-intercept form   **point-slope form**   **standard form**

1. A linear equation in the form $y - y_1 = m(x - x_1)$ is written in

   _____.

2. A linear equation in the form $Ax + By = C$ is written in

   _____.

3. A linear equation in the form $y = mx + b$ is written in

   _____.

Objective 1   Use point-slope form to write an equation of a line.

Video Examples

Review this example for Objective 1:

1b. Write an equation of each line. Give the final answer in slope-intercept form.

The line passing through $(5, -3)$ with slope $\dfrac{5}{4}$.

$$y - y_1 = m(x - x_1)$$
$$y - (-3) = \frac{5}{4}(x - 5)$$
$$y + 3 = \frac{5}{4}x - \frac{25}{4}$$
$$y = \frac{5}{4}x - \frac{37}{4}$$

Now Try:

1b. Write an equation of each line. Give the final answer in slope-intercept form.

The line passing through $(6, 11)$ with slope $-\dfrac{2}{3}$.

_____

Objective 1 Practice Exercises

For extra help, see Example 1 on page 266 of your text.

Write an equation of each line. Give the final answer in slope-intercept form.

1. The line passing through $(8,-7)$ with slope $\frac{1}{4}$.

    1. _____

2. The line passing through $(-3, 6)$ with slope -4.

    2. _____

3. The line passing through $(5, 2)$ with slope 2.

    3. _____

Objective 2   Write an equation of a line by using two points on the line.

Video Examples

Review this example for Objective 2:

2. Write the equation of the line passing through the point $(6, 8)$ and $(-3, 5)$. Give the final answer in slope-intercept form and then in standard form.

    First, find the slope of the line.
    $(x_1, y_1) = (6,\ 8)$ and $(x_2, y_2) = (-3,\ 5)$

    $$\text{slope } m = \frac{y_2 - y_1}{x_2 - x_1} = \frac{5-8}{-3-6} = \frac{-3}{-9} = \frac{1}{3}$$

    Now use (x_1, y_1), here $(6, 8)$ and point-slope form.

Now Try:

2. Write the equation of the line passing through the point $(7, 15)$ and $(15, 9)$. Give the final answer in slope-intercept form and then in standard form.

    _____

$$y - y_1 = m(x - x_1)$$

$$y - 8 = \frac{1}{3}(x - 6)$$

$$y - 8 = \frac{1}{3}x - 2$$

$$y = \frac{1}{3}x + 6 \quad \text{Slope-intercept form}$$

$$3y = x + 18$$

$$-x + 3y = 18$$

$$x - 3y = -18 \qquad \text{Standard form}$$

Objective 2 Practice Exercises

For extra help, see Example 2 on page 267 of your text.

Write an equation for the line passing through each pair of points. Write the equations in standard form.

4. $(-2, 1)$ and $(3, 11)$ 4. _____

5. $(2, 3)$ and $(-2, -3)$ 5. _____

6. $(3, -4)$ and $(2, 7)$ 6. _____

Objective 3   Find an equation of a line that fits a data set.

Video Examples

Review this example for Objective 3:

3.   The table shows the number of internet users in the world from 1998 to 2005, where year 0 represents 1998.

Year	Number of Internet Users (millions)
0	147
2	361
4	587
6	817
8	1093

Plot the data and find an equation that approximates it.

Letting y represent the number of internet users in year x, we plot the data.

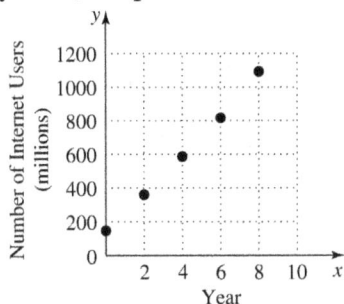

The points appear to lie approximately in a straight line. To find an equation of the line, we choose the ordered pairs (0, 147) and (8, 1093) from the table and find the slope of the line through these points.

$$(x_1, y_1) = (0,\ 147) \text{ and } (x_2, y_2) = (8,\ 1093)$$

$$\text{slope } m = \frac{y_2 - y_1}{x_2 - x_1} = \frac{1093 - 147}{8 - 0} = \frac{946}{8}$$

$$= 118.25$$

Use the slope, 118.25, and the point (0, 147) in slope-intercept form.

$$y = mx + b$$

$$147 = 118.25(0) + b$$

$$147 = b$$

Thus, $m = 118.25$ and $b = 147$, so the equation of the line is $y = 118.25x + 147$.

Now Try:

3.   The table shows the average annual telephone expenditures for residential telephones from 2001 to 2006, where year 0 represents 2001.

Year	Annual Telephone Expenditures
0	$686
2	$620
3	$592
4	$570
5	$542

Plot the data and find an equation that approximates it.

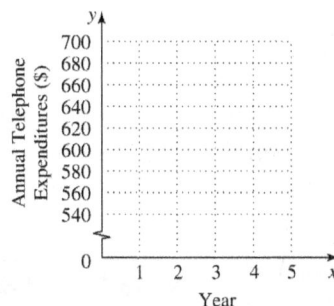

_____

Name: _____   Date: _____

Instructor: _____   Section: _____

Objective 3 Practice Exercises

For extra help, see Example 8 on page 260 of your text.

Plot the data and find an equation that approximates it.

7. The table shows the U.S. municipal solid waste recycling percents since 1985, where year 0 represents 1985.

Year	Recycling Percent
0	10.1
5	16.2
10	26.0
15	29.1
20	32.5

7.

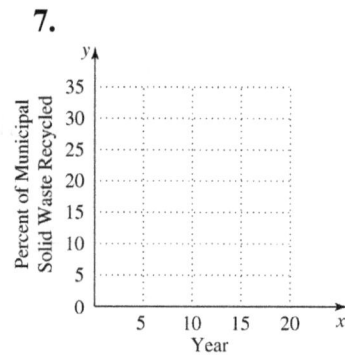

_____

8. The table shows the approximate consumer expenditures for food in the U.S. in billions of dollars for selected years, where year 0 represents 1985.

Year	Food Expenditures (billions of dollars)
0	233
5	298
10	343
15	417
20	515

8.

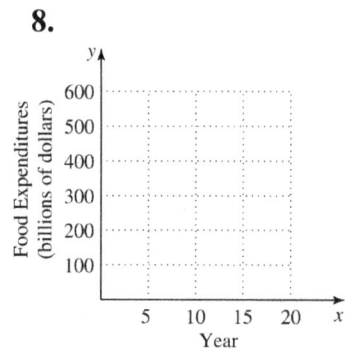

_____

Chapter 3 GRAPHS OF LINEAR EQUATIONS AND INEQUALITIES IN TWO VARIABLES

3.6   Graphing Linear Inequalities in Two Variables

Learning Objectives
1   Graph linear inequalities in two variables.
2   Graph an inequality with a boundary line through the origin.

Key Terms

Use the vocabulary terms listed below to complete each statement in exercises 1–2.

linear inequality in two variables   boundary line

1.   In the graph of a linear inequality, the _____ separates the region that satisfies the inequality from the region that does not satisfy the inequality.

2.   An inequality that can be written in the form $Ax + By < C$, $Ax + By > C$, $Ax + By \leq C$, or $Ax + By \geq C$ is called a _____.

Objective 1   Graph linear inequalities in two variables.

Video Examples

Review these examples for Objective 1:

2.   Graph $2x + 5y > -10$.

This equation does not include equality. Therefore, the points on the line $2x + 5y = -10$ do not belong to the graph. However, the line still serves as a boundary for two regions.

To graph the inequality, first graph the equation $2x + 5y > -10$. Use a dashed line to show that the points on the line are not solutions of the inequality $2x + 5y > -10$. Then choose a test point to see which region satisfies the inequality.

$$2x + 5y > -10$$
$$2(0) + 5(0) \overset{?}{>} -10$$
$$0 + 0 \overset{?}{>} -10$$
$$0 > -10 \text{ True}$$

Since $0 > -10$ is true, the graph of the inequality is the region that contains $(0, 0)$. Shade that region.

Now Try:

2.   Graph $5x + 4y > 20$.

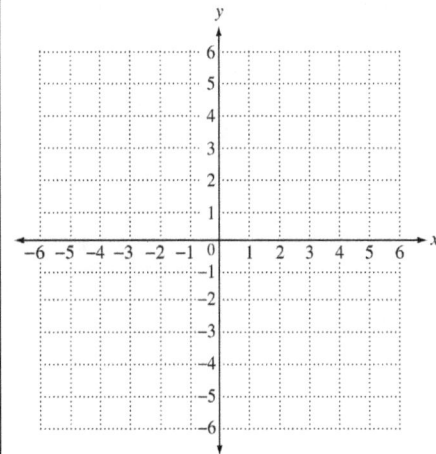

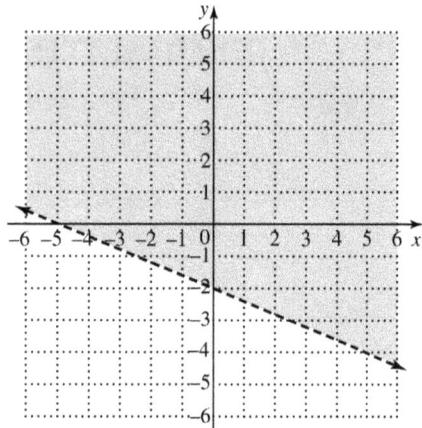

3. Graph $x - 4 \le -1$.

First, solve the inequality for x.

$x \le 3$

Now graph the line $x = 3$, a vertical line through the point (3, 0). Use a solid line, and choose (0, 0) as a test point.

$0 \le 3$ True

Because $0 \le 3$ is true, we shade the region containing (0, 0).

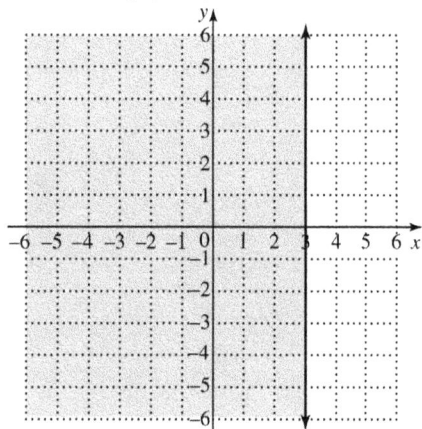

3. Graph $y \ge -1$.

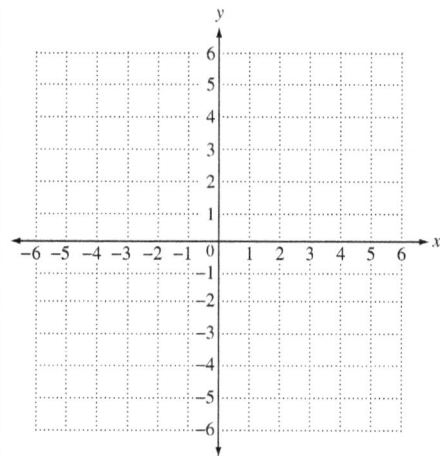

Name: Date:
Instructor: Section:

Objective 1 Practice Exercises

For extra help, see Examples 1–3 on pages 277–279 of your text.

Graph each linear inequality.

1. $y \geq x - 1$

1.

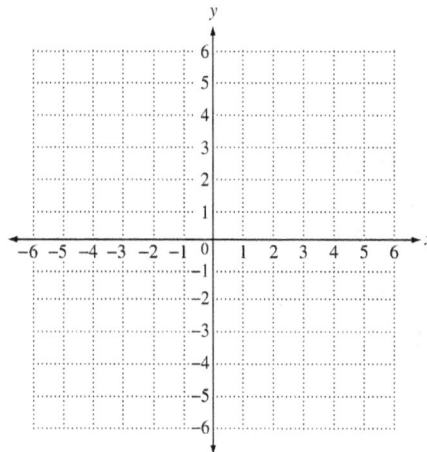

2. $y > -x + 2$

2.

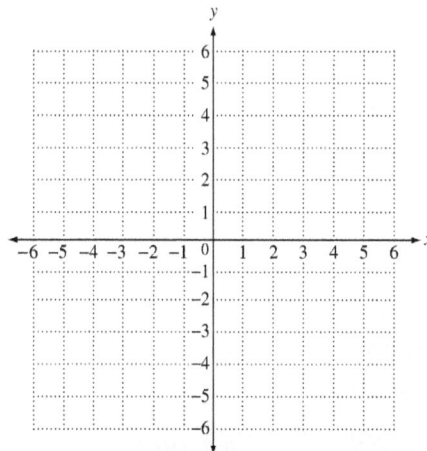

3. $3x - 4y - 12 > 0$

3.

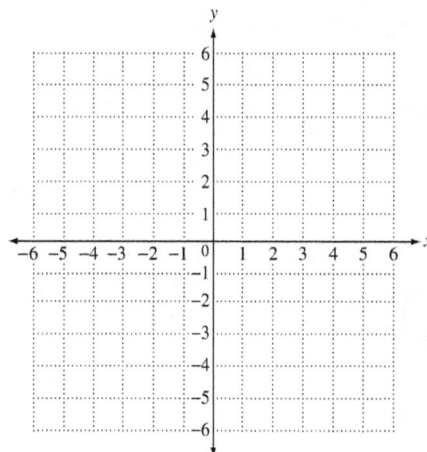

Objective 2   Graph an inequality with a boundary line through the origin.

Video Examples

Review this example for Objective 2:

4.   Graph $y \geq 3x$.

We graph $y = 3x$ using a solid line through $(0, 0)$, $(1, 3)$ and $(2, 6)$. Because $(0, 0)$ is on the line $y \geq 3x$, it cannot be used as a test point. Instead we choose a test point off the line, say $(3, 0)$.

$$0 \overset{?}{\geq} 3(3)$$

$$0 \geq 9 \quad \text{False}$$

Because $0 \geq 9$ is false, shade the other region.

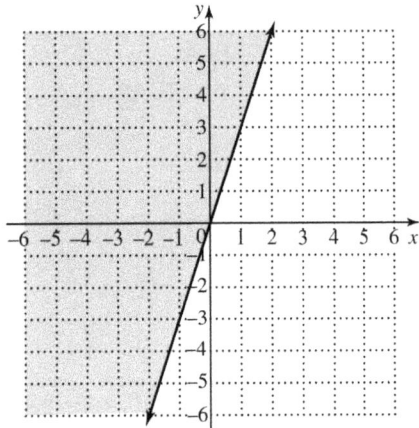

Now Try:

4.   Graph $y \geq x$.

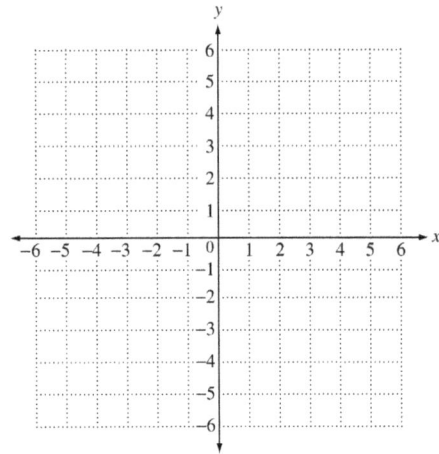

Objective 2 Practice Exercises

For extra help, see Example 4 on page 279 of your text.

Graph each linear inequality.

4. $y \leq \dfrac{2}{5}x$

4.

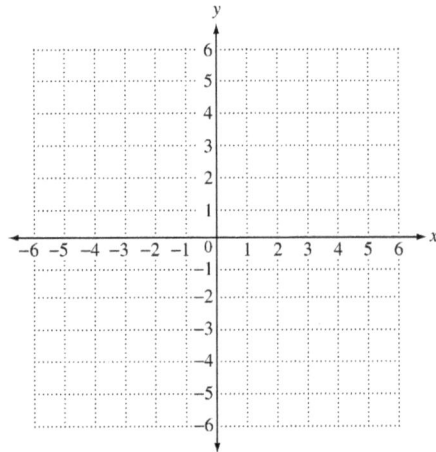

5. $y \geq \dfrac{1}{3}x$

5.

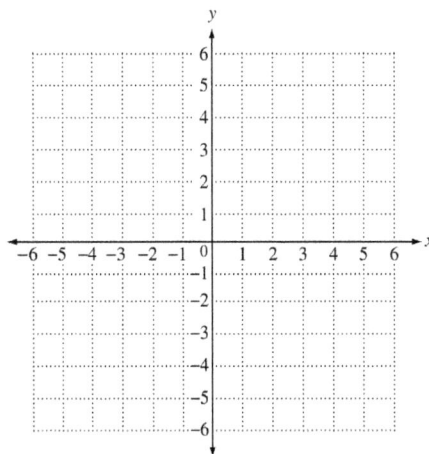

6. $x < -2y$

6.

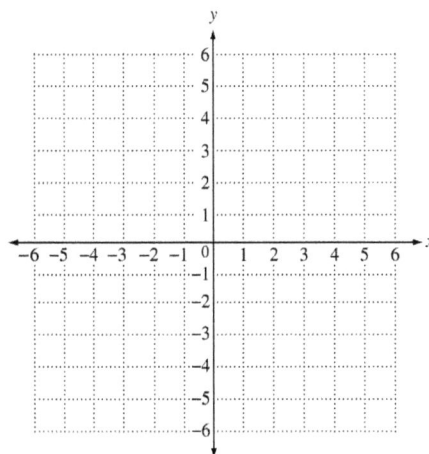

Chapter 4 SYSTEMS OF LINEAR EQUATIONS AND INEQUALITIES

4.1   Solving Systems of Linear Equations by Graphing

Learning Objectives
1   Decide whether a given ordered pair is a solution of a system.
2   Solve linear systems by graphing.
3   Solve special systems by graphing.
4   Identify special systems without graphing.

Key Terms

Use the vocabulary terms listed below to complete each statement in exercises 1–7.

system of linear equations   **solution of a system**

solution set of a system   **consistent system**

inconsistent system   **independent equations**

dependent equations

1.   Equations of a system that have different graphs are called
      _____.

2.   A system of equations with at least one solution is a
      _____.

3.   The set of all ordered pairs that are solutions of a system is the
      _____.

4.   The _____ of linear equations is an ordered
      pair that makes all the equations of the system true at the same time.

5.   Equations of a system that have the same graph (because they are different forms of
      the same equation) are called _____.

6.   A system with no solution is called a(n) _____.

7.   A(n) _____ consists of two or more linear
      equations with the same variables.

Objective 1   Decide whether a given ordered pair is a solution of a system.

Video Examples

Review this example for Objective 1:

1b. Determine whether the ordered pair (5, –2) is a solution of the system.

$$4x + 5y = 10$$

$$3x + 8y = 6$$

Again, substitute 5 for x and –2 for y in each equation.

$4x + 5y = 10$	$3x + 8y = 6$
$4(5) + 5(-2) \overset{?}{=} 10$	$3(5) + 8(-2) \overset{?}{=} 6$
$20 - 10 \overset{?}{=} 10$	$15 - 16 \overset{?}{=} 6$
$10 = 10$ True	False $-1 = 6$

The ordered pair (5, –2) is not a solution of this system because it does not satisfy the second equation.

Now Try:

1b. Determine whether the ordered pair (6, 5) is a solution of the system.

$$5x - 6y = 0$$

$$6x + 5y = 50$$

_____

Objective 1 Practice Exercises

For extra help, see Example 1 on page 296 of your text.

Decide whether the given ordered pair is a solution of the given system.

1. $(2, -4)$

  $2x + 3y = 6$

  $3x - 2y = 14$

1. _____

2. $(-3, -1)$

  $5x - 3y = -12$

  $2x + 3y = -9$

2. _____

3. $(4, 0)$

  $4x + 3y = 16$

  $x - 4y = -4$

3. _____

Objective 2   Solve linear systems by graphing.

Video Examples

Review this example for Objective 2:

2.   Solve the system of equation by graphing both equations on the same axes.

$$6x - 5y = 4$$

$$2x - 5y = 8$$

Graph these equations by plotting several points for each line. To find the *x*-intercept, let $y = 0$. To find the *y*-intercept, let $x = 0$.

The tables show the intercepts and a check point for each graph.

$$6x - 5y = 4$$

x	y
0	$-\dfrac{4}{5}$
$\dfrac{2}{3}$	0
4	4

$$2x - 5y = 8$$

x	y
0	$-\dfrac{8}{5}$
4	0
-6	-4

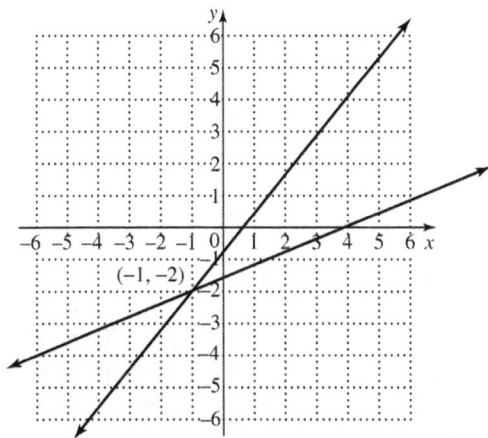

The lines suggest that the graphs intersect at the point $(-1, -2)$. We check by substituting -1 for *x* and -2 for *y* in both equations.

$$6x - 5y = 4$$
$$6(-1) - 5(-2) \overset{?}{=} 4$$
$$-6 + 10 \overset{?}{=} 4$$
$$4 = 4 \quad \text{True}$$

$$2x - 5y = 8$$
$$2(-1) - 5(-2) \overset{?}{=} 8$$
$$-2 + 10 \overset{?}{=} 8$$
$$\text{True} \quad 8 = 8$$

Because $(-1, -2)$ satisfies both equations, the solution set of this system is $\{(-1, -2)\}$.

Now Try:

2.   Solve the system of equation by graphing both equations on the same axes.

$$3x - y = -7$$

$$2x + y = -3$$

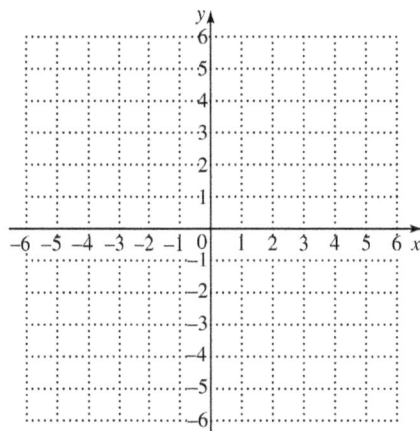

Name: Date:

Instructor: Section:

Objective 2 Practice Exercises

For extra help, see Example 2 on page 297 of your text.

Solve each system by graphing both equations on the same axes.

4. $x - 2y = 6$ 4.

       $2x + y = 2$

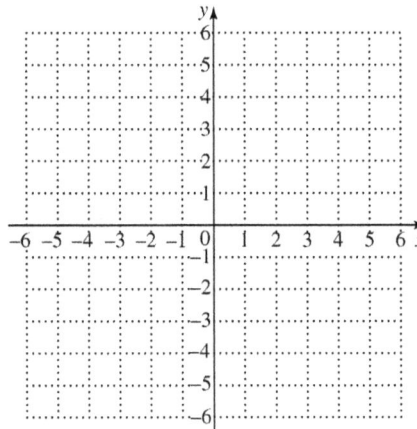

_____

5. $2x = y$ 5.

       $5x + 3y = 0$

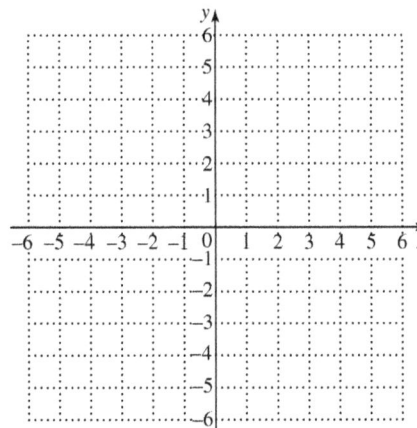

_____

6. $3x + 2 = y$

$2x - y = 0$

6.

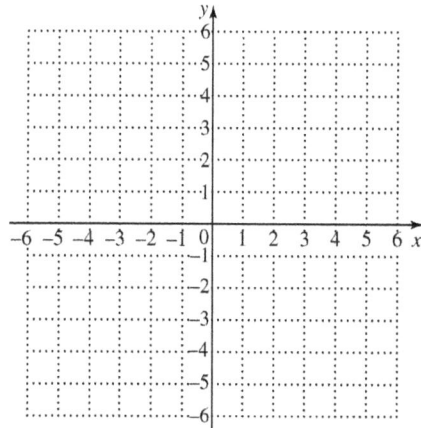

_____

Objective 3   Solve special systems by graphing.

Video Examples

Review these examples for Objective 3:

3. Solve each system by graphing.

a. $x - y = 1$

$x - y = -1$

The graphs of these two equations are parallel and have no points in common. There is no solution for this system. The solution set is $\varnothing$.

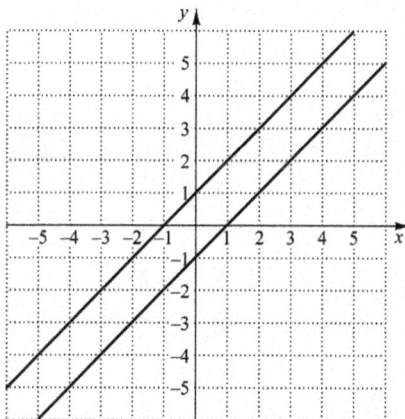

Now Try:

3. Solve each system by graphing.

a. $4x - 2y = 8$

$6x - 3y = 12$

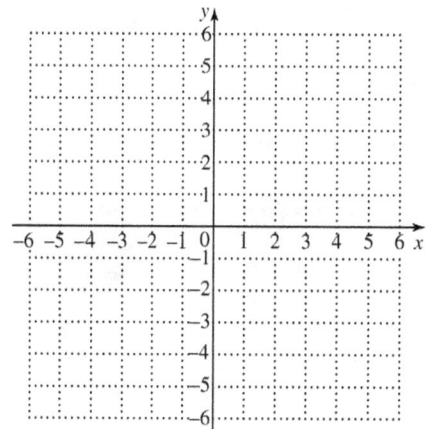

_____

b. $3x - y = 0$

$\qquad 2y = 6x$

The graphs of these two equations are the same line.

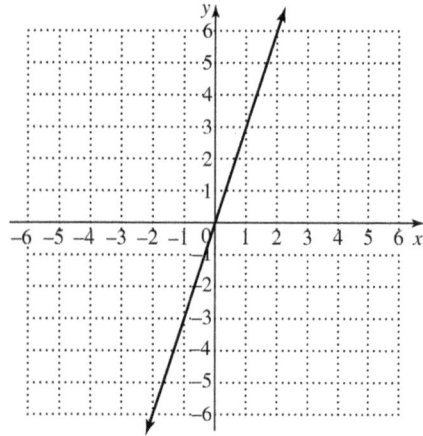

In this case, every point on the line is a solution of the system, and the solution set contains an infinite number of ordered pairs, each of which satisfies both equations of the system. We write the solution set as

$\qquad \{(x, y) \mid 3x - y = 0\}$.

b. $x - 3y = 6$

$\qquad x - 3y = 4$

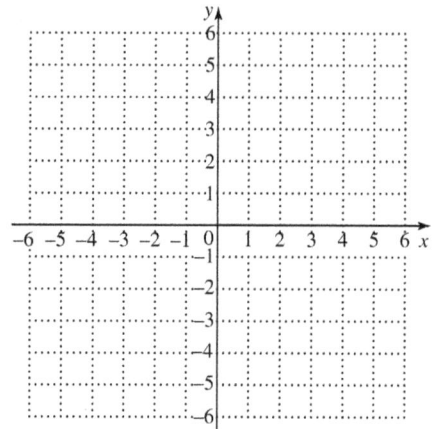

_____

Objective 3 Practice Exercises

For extra help, see Example 3 on pages 298–299 of your text.

*Solve each system of equations by graphing both equations on the same axes. If the two equations produce parallel lines, write **no solution**. If the two equations produce the same line, write **infinite number of solutions**.*

7. $\quad 8x + 4y = -1$

$\qquad 4x + 2y = 3$

7.

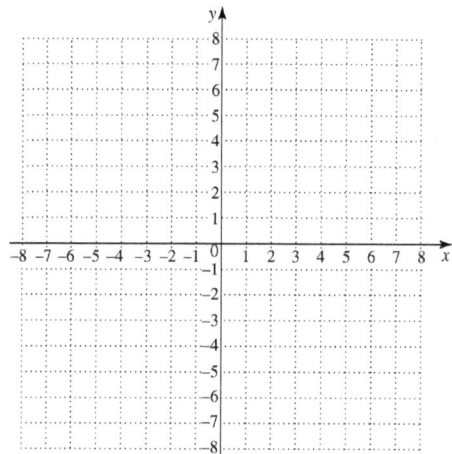

_____

8. $-3x + 2y = 6$

  $-6x + 4y = 12$

8.

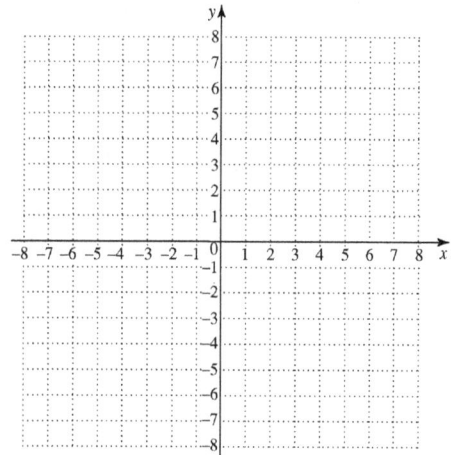

Objective 4   Identify special systems without graphing.

Video Examples

Review this example for Objective 4:

4c. Describe the system without graphing. State the number of solutions.

$3x - 4y = 12$

$2x - 3y = 6$

Write each equation in slope-intercept form.

$3x - 4y = 12$	$2x - 3y = 6$
$-4y = -3x + 12$	$-3y = -2x + 6$
$y = \dfrac{3}{4}x - 3$	$y = \dfrac{2}{3}x - 2$

The graphs are neither parallel nor the same line, since the slopes are different. This system has exactly one solution.

Now Try:

4c. Describe the system without graphing. State the number of solutions.

$x - 6y = 4$

$6x - y = 9$

Objective 4 Practice Exercises

For extra help, see Example 4 on pages 300–301 of your text.

Without graphing, answer the following equations for each linear system.
(a) Is the system inconsistent, are the equations dependent, or neither?
(b) Is the graph a pair of intersecting lines, a pair of parallel lines, or one line?
(c) Does the system have one solution, no solution, or an infinite number of solutions?

9. $y = 2x + 1$

           $3x - y = 7$

9. (a)_____

    (b)_____

    (c)_____

10. $-2x + y = 4$

        $-4x + 2y = -2$

10. (a)_____

     (b)_____

     (c)_____

11. $4x + 3y = 12$

        $-12x = -36 + 9y$

11. (a)_____

     (b)_____

     (c)_____

Chapter 4 SYSTEMS OF LINEAR EQUATIONS AND INEQUALITIES

4.2   Solving Systems of Linear Equations by Substitution

Learning Objectives
1   Solve linear systems by substitution.
2   Solve special systems by substitution.
3   Solve linear systems with fractions and decimals.

Key Terms

Use the vocabulary terms listed below to complete each statement in exercises 1−4.

substitution   **ordered pair**   **inconsistent system**

dependent system

1.   The solution of a linear system of equations is written as a(n) _____.

2.   When one expression is replaced by another, _____ is being used.

3.   A system of equations in which all solutions of the first equation are also solutions of the second equation is a(n) _____.

4.   A system of equations that has no common solution is called a(n)
_____.

Objective 1   Solve linear systems by substitution.

Video Examples

Review these examples for Objective 1:
1.   Solve the system by the substitution method.
$$2x + 5y = 22 \quad (1)$$
$$y = 4x \quad (2)$$

Equation (2) is already solved for y. We substitute $4x$ for y in equation (1).
$$2x + 5y = 22$$
$$2x + 5(4x) = 22$$
$$2x + 20x = 22$$
$$22x = 22$$
$$x = 1$$
Find the value of y by substituting 1 for x in either equation. We use equation (2).
$$y = 4x$$
$$y = 4(1) = 4$$
We check the solution (1, 4) by substituting 1 for

Now Try:
1.   Solve the system by the substitution method.
$$x + y = 7$$
$$y = 6x$$

_____

x and 4 for y in both equations.

$$2x + 5y = 22 \qquad\qquad y = 4x$$
$$2(1) + 5(4) \overset{?}{=} 22 \qquad 4 \overset{?}{=} 4(1)$$
$$2 + 20 \overset{?}{=} 22 \qquad \text{True } 4 = 4$$
$$22 = 22 \text{ True}$$

Since $(1, 4)$ satisfies both equations, the solution set of the system is $\{(1, 4)\}$.

2. Solve the system by the substitution method.

$$4x + 5y = 13 \qquad (1)$$
$$x = -y + 2 \qquad (2)$$

Equation (2) gives x in terms of y. We substitute $-y + 2$ for x in equation (1).

$$4x + 5y = 13$$
$$4(-y + 2) + 5y = 13$$
$$-4y + 8 + 5y = 13$$
$$y + 8 = 13$$
$$y = 5$$

Find the value of x by substituting 5 for y in either equation. We use equation (2).

$$x = -y + 2$$
$$x = -5 + 2 = -3$$

We check the solution $(-3, 5)$ by substituting -3 for x and 5 for y in both equations.

$$4x + 5y = 13 \qquad\qquad x = -y + 2$$
$$4(-3) + 5(5) \overset{?}{=} 13 \qquad -3 \overset{?}{=} -5 + 2$$
$$-12 + 25 \overset{?}{=} 13 \qquad \text{True } -3 = -3$$
$$13 = 13 \text{ True}$$

Both results are true, so the solution set of the system is $\{(-3, 5)\}$.

3. Solve the system by the substitution method.

$$4x = 5 - y \qquad (1)$$
$$7x + 3y = 15 \qquad (2)$$

Step 1 Solve one of the equations for x or y. Solve equation (1) for y to avoid fractions.

$$4x = 5 - y$$
$$y + 4x = 5$$
$$y = -4x + 5$$

2. Solve the system by the substitution method.

$$2x + 3y = 6$$
$$x = 5 - y$$

_____

3. Solve the system by the substitution method.

$$2x + 7y = 2$$
$$3y = 2 - x$$

_____

Step 2 Now substitute $-4x+5$ for y in equation (2).

$$7x+3y=15$$

$$7x+3(-4x+5)=15$$

Step 3 Solve the equation from Step 2.

$$7x-12x+15=15$$

$$-5x+15=15$$

$$-5x=0$$

$$x=0$$

Step 4 Equation (1) solved for y is $y=-4x+5$. Substitute 0 for x.

$$y=-4(0)+5=5$$

Step 5 Check that $(0, 5)$ is the solution.

$$
\begin{array}{c|c}
4x=5-y & 7x+3y=15 \\
4(0)\overset{?}{=}5-5 & 7(0)+3(5)\overset{?}{=}15 \\
0=0 \ \ \text{True} & \text{True} \ \ 15=15
\end{array}
$$

Since both results are true, the solution set of the system is $\{(0, 5)\}$.

Objective 1 Practice Exercises

For extra help, see Examples 1–3 on pages 306–308 of your text.

Solve each system by the substitution method. Check each solution.

1. $3x+2y=14$ 1. _____

$$y=x+2$$

2. $x+\ y=9$ 2. _____

$$5x-2y=-4$$

3. $3x - 21 = y$ **3.** _____

         $y + 2x = -1$

Objective 2   Solve special systems by substitution.

Video Examples

Review these examples for Objective 2:

4.   Use substitution to solve the system.

$$x = 7 - 3y \quad (1)$$
$$5x + 15y = 1 \qquad (2)$$

Because equation (1) is already solved for x, we substitute $7 - 3y$ for x in equation (2).

$$5x + 15y = 1$$
$$5(7 - 3y) + 15y = 1$$
$$35 - 15y + 15y = 1$$
$$35 = 1 \quad \text{False}$$

A false result, here $35 = 1$, means that the equations in the system have graphs that are parallel lines. The system is inconsistent and has no solution, so the solution set is $\varnothing$.

5.   Use substitution to solve the system.

$$14x - 7y = 21 \quad (1)$$
$$-2x + y = -3 \quad (2)$$

Begin by solving equation (2) for y to get $y = 2x - 3$. Substitute $2x - 3$ for y in equation (1).

$$14x - 7y = 21$$
$$14x - 7(2x - 3) = 21$$
$$14x - 14x + 21 = 21$$
$$0 = 0$$

This true result means that every solution of one equation is also a solution of the other, so the system has an infinite number of solutions. The solution set is $\{(x, y) | 14x - 7y = 21\}$.

Now Try:

4.   Use substitution to solve the system.

$$5x - 10y = 8$$
$$x = 2y + 5$$

_____

5.   Use substitution to solve the system.

$$5x + 4y = 20$$
$$-10x + 40 = 8y$$

_____

Name: _____   Date: _____

Instructor: _____   Section: _____

Objective 2 Practice Exercises

For extra help, see Examples 4–5 on pages 308–309 of your text.

Solve each system by the substitution method. Use set-builder notation for dependent equations.

4. $y = -\dfrac{1}{3}x + 5$

       $3y + x = -9$

       4. _____

5. $\dfrac{1}{2}x + 3 = y$

       $6 = -x + 2y$

       5. _____

6. $4x + 3y = 2$

       $8x + 6y = 6$

       6. _____

Objective 3   Solve linear systems with fractions and decimals.

Video Examples

Review this example for Objective 3:

6.   Solve the system by the substitution method.

$$\frac{1}{2}x - y = 3 \quad (1)$$

$$\frac{1}{5}x + \frac{1}{2}y = \frac{3}{10} \quad (2)$$

Clear equation (1) of fractions by multiplying each side by 2.

$$2\left(\frac{1}{2}x - y\right) = 2(3)$$

$$2\left(\frac{1}{2}x\right) - 2y = 2(3)$$

$$x - 2y = 6$$

Clear equation (2) of fractions by multiplying each side by 10.

$$10\left(\frac{1}{5}x + \frac{1}{2}y\right) = 10\left(\frac{3}{10}\right)$$

$$10\left(\frac{1}{5}x\right) + 10\left(\frac{1}{2}y\right) = 10\left(\frac{3}{10}\right)$$

$$2x + 5y = 3$$

The given system of equations has been simplified to an equivalent system.

$$x - 2y = 6 \quad (3)$$

$$2x + 5y = 3 \quad (4)$$

To solve the system by substitution, solve equation (3) for x.

$$x - 2y = 6$$

$$x = 2y + 6$$

Now substitute the result for x in equation (4).

$$2x + 5y = 3$$

$$2(2y + 6) + 5y = 3$$

$$4y + 12 + 5y = 3$$

$$9y + 12 = 3$$

$$9y = -9$$

$$y = -1$$

Substitute -1 for y in $x = 2y + 6$ (equation (3) solved for x).

$$x = 2(-1) + 6 = 4$$

Check $(4, -1)$ in both of the original equations. The solution set is $\{(4, -1)\}$.

Now Try:

6.   Solve the system by the substitution method.

$$x + \frac{1}{2}y = \frac{1}{2}$$

$$\frac{1}{2}x + \frac{1}{5}y = 0$$

_____

Objective 3 Practice Exercises

For extra help, see Examples 6–7 on pages 310–311 of your text.

Solve each system by the substitution method. Check each solution.

7. $\frac{5}{4}x - y = -\frac{1}{4}$

     $-\frac{7}{8}x + \frac{5}{8}y = 1$

7. _____

8. $\frac{1}{4}x + \frac{3}{8}y = -3$

     $\frac{5}{6}x - \frac{3}{7}y = -10$

8. _____

9. $0.6x + 0.8y = 1$

     $0.4y = 0.5 - 0.3x$

9. _____

Chapter 4 SYSTEMS OF LINEAR EQUATIONS AND INEQUALITIES

4.3   Solving Systems of Linear Equations by Elimination

Learning Objectives
1   Solve linear systems by elimination.
2   Multiply when using the elimination method.
3   Use an alternative method to find the second value in a solution.
4   Solve special systems by elimination.

Key Terms

Use the vocabulary terms listed below to complete each statement in exercises 1−3.

addition property of equality   elimination method   substitution

1.   Using the addition property to solve a system of equations is called the

_____.

2.   The _____ states that the same added quantity to each side of an equation results in equal sums.

3.   _____ is being used when one expression is replaced by another.

Objective 1   Solve linear systems by elimination.

Video Examples

Review this example for Objective 1:	**Now Try:**
1. Use the elimination method to solve the system.	1. Use the elimination method to solve the system.

Review this example for Objective 1:

1.   Use the elimination method to solve the system.
$$x + y = 6 \quad (1)$$
$$-x + y = 4 \quad (2)$$

Add the equations vertically.
$$x + y = 6 \quad (1)$$
$$\underline{-x + y = 4} \quad (2)$$
$$2y = 10$$
$$y = 5$$

To find the x-value, substitute 5 for y in either of the two equations of the system. We choose equation (1).
$$x + y = 6$$
$$x + 5 = 6$$
$$x = 1$$

Check the solution $(1, 5)$, by substituting 1 for x and 5 for y in both equations of the given system.

Now Try:

1.   Use the elimination method to solve the system.
$$x + y = 11$$
$$x - y = 5$$

_____

$$x + y = 6 \qquad\qquad -x + y = 4$$
$$1 + 5 \stackrel{?}{=} 6 \qquad\qquad -1 + 5 \stackrel{?}{=} 4$$
$$\qquad 6 = 6 \ \text{True} \qquad \text{True} \ 4 = 4$$

Since both results are true, the solution set of the system is $\{(1, 5)\}$.

Objective 1 Practice Exercises

For extra help, see Examples 1–2 on pages 314–316 of your text.

Solve each system by the elimination method. Check your answers.

1. $x - 4y = -4$

       $-x + y = -5$

1. _____

2. $2x - y = 10$

       $3x + y = 10$

2. _____

3. $x - 3y = 5$

       $-x + 4y = -5$

3. _____

Name: _____ Date: _____

Instructor: _____ Section: _____

Objective 2   Multiply when using the elimination method.

Video Examples

Review these examples for Objective 2:

4. Solve the system.

$$3x + 8y = -2 \quad (1)$$

$$2x + 7y = 2 \quad (2)$$

To eliminate x, multiply equation (1) by 2 and multiply equation (2) by –3. Then add.

$$6x + 16y = -4$$

$$\underline{-6x - 21y = -6}$$

$$-5y = -10$$

$$y = 2$$

Substituting 2 for y in either equation (1) or (2) gives $x = -6$. Check that the solution set of the system is {(–6, 2)}.

3. Solve the system.

$$4x + 2y = 0 \quad (1)$$

$$x + 3y = 5 \quad (2)$$

Step 1 The equations are already written in $Ax + By = C$ form.

Step 2 Adding the two equations gives $5x + 5y = 5$, which does not eliminate either variable. However, multiplying equation (2) by –4 and then adding will eliminate the variable x.

Step 3 Add the two equations.

$$4x + 2y = 0$$

$$\underline{-4x - 12y = -20}$$

$$-10y = -20$$

Step 4 Solve. $y = 2$

Step 5 Find the value of x by substituting 2 for y in either of the original equations. We choose equation (2).

$$x + 3y = 5$$

$$x + 3(2) = 5$$

$$x = -1$$

Step 6 A check of the ordered pair (–1, 2) by substituting $x = -1$ and $y = 2$ in both of the original equations shows that the solution set of the system is {(–1, 2)}.

Now Try:

4. Solve the system.

$$3x + 4y = 24$$

$$4x + 3y = 11$$

_____

3. Solve the system.

$$2x - 3y = 11$$

$$x + 4y = -11$$

_____

Name: Date:

Instructor: Section:

Objective 2 Practice Exercises

For extra help, see Examples 3–4 on pages 316–317 of your text.

Solve each system by the elimination method. Check your answers.

4. $6x + 7y = 10$ 4. _____

     $2x - 3y = 14$

5. $8x + 6y = 10$ 5. _____

     $4x - y = 1$

6. $6x + y = 1$ 6. _____

     $3x - 4y = 23$

169

Name: Date:

Instructor: Section:

Objective 3 Use an alternative method to find the second value in a solution.

Video Examples

Review this example for Objective 3:	**Now Try:**
5. Solve the system.	**5.** Solve the system.

5. Solve the system.

$$6x = 7 - 3y \quad (1)$$

$$8x - 5y = 5 \quad (2)$$

Write equation (1) in standard form.

$$6x + 3y = 7 \quad (3)$$

$$8x - 5y = 5 \quad (4)$$

One way to proceed is to eliminate y by multiplying each side of equation (3) by 5 and each side of equation (4) by 3, and then adding.

$$30x + 15y = 35$$

$$\underline{24x - 15y = 15}$$

$$54x \qquad = 50$$

$$x = \frac{50}{54} \text{ or } \frac{25}{27}$$

Substituting $\frac{25}{27}$ for x in one of the given

equations would give y, but the arithmetic would be complicated. Instead, solve for y by starting again with the original equations in standard form, and eliminating x. Multiply equation (3) by 4 and equation (4) by –3.

$$24x + 12y = 28$$

$$\underline{-24x + 15y = -15}$$

$$27y = 13$$

$$y = \frac{13}{27}$$

The solution set is $\left\{ \left(\frac{25}{27}, \frac{13}{27} \right) \right\}$.

Now Try:

5. Solve the system.

$$8x = 5y + 1$$

$$6x - 8y = -2$$

_____

Objective 3 Practice Exercises

For extra help, see Example 5 on pages 317–318 of your text.

Solve each system by the elimination method. Check your answers.

7. $\quad 4x - 3y - 20 = 0$

$\quad\quad 6x + 5y + 8 = 0$

7. _____

8. $6x = 16 - 7y$

  $4x = 3y + 26$

8. _____

9. $2x = 14 + 4y$

  $6y = -5x + 3$

9. _____

Objective 4 Solve special systems by elimination.

Video Examples

Review these examples for Objective 4:

6. Solve each system by the elimination method.

  a. $5x + 10y = 9$ (1)

    $3x + 6y = 8$ (2)

Multiply each side of equation (1) by 3 and each side of equation (2) by –5.

$$15x + 30y = 27$$
$$\underline{-15x - 30y = -40}$$
$$0 = -13 \text{ False}$$

The false statement $0 = -13$ indicates that the system has solution set $\varnothing$.

Now Try:

6. Solve each system by the elimination method.

  a. $2x + 6y = 5$

    $5x + 15y = 8$

    _____

b. $7x + y = 9$ (1)

    $-14x - 2y = -18$ (2)

Multiply each side of equation (1) by 2.

    $14x + 2y = 18$

    $\underline{-14x - 2y = -18}$

           $0 = 0$ True

A true statement occurs when the equations are equivalent. This indicates that every solution of one equation is also a solution of the other. The solution set is $\{(x, y) \mid 7x + y = 9\}$.

b. $9x - 7y = 5$

    $18x = 14y + 10$

_____

Objective 4 Practice Exercises

For extra help, see Example 5 on page 318 of your text.

Solve each system by the elimination method. Use set-builder notation for dependent equations. Check your answers.

10. $12x - 8y = 3$

       $6x - 4y = 6$

10. _____

11. $2x + 4y = -6$

       $-x - 2y = 3$

11. _____

12. $15x + 6y = 9$

       $10x + 4y = 18$

12. _____

Chapter 4 SYSTEMS OF LINEAR EQUATIONS AND INEQUALITIES

4.4   Applications of Linear Systems

Learning Objectives
1   Solve problems about unknown numbers.
2   Solve problems about quantities and their costs.
3   Solve problems about mixtures.
4   Solve problems about distance, rate (or speed), and time.

Key Terms

Use the vocabulary terms listed below to complete each statement in exercises 1−2.

system of linear equations $d = rt$

1. The formula that relates distance, rate, and time is _____.

2. A _____ consists of at least two linear equations with different variables.

Objective 1   Solve problems about unknown numbers.

Video Examples

Review this example for Objective 1:

1. Two towns have a combined population of 9045. There are 2249 more people living in one than in the other. Find the population in each town.

    Step1 Read the problem carefully. We are to find the population of each town.

    Step 2 Assign variables. Let x = the population of the larger town, and y = the population of smaller town.

    Step 3 Write two equations. There are 2249 more people living in one town. The total population is 9045.

    $x = y + 2249$ (1)

    $x + y = 9045$ (2)

    Step 4 Solve the system. We use substitution. Substitute $y + 2249$ for x in equation (2).

Now Try:

1. A rope 82 centimeters long is cut into two pieces with one piece four more than twice as long as the other. Find the length of each piece.

    _____

$$x + y = 9045$$
$$(y + 2249) + y = 9045$$
$$2249 + 2y = 9045$$
$$2y = 6796$$
$$y = 3398$$

To find x, substitute 3398 into either original equation. We use equation (1).

$$x = 3398 + 2249 = 5647$$

Step 5 State the answer. The population of the larger town is 5647. The population of the smaller town is 3398.

Step 6 Check the answer in the original problem. $3398 + 2249 = 5647$ and $5647 + 3398 = 9045$ The answers check.

Objective 1 Practice Exercises

For extra help, see Example 1 on pages 324–325 of your text.

Write a system of equations for each problem, then solve the problem.

1. The difference between two numbers is 14. If two times the smaller is added to one-half the larger, the result is 52. Find the numbers.

    1.
    larger number _____

    smaller number _____

2. There are a total of 49 students in the two second grade classes at Jefferson School. If Carla has 7 more students in her class than Linda, find the number of students in each class.

    2.
    Carla's class _____

    Linda's class _____

3.   The perimeter of a rectangular room is 50 feet. The   **3.**
     length is three feet greater than the width. Find the   length _____
     dimensions of the rectangle.   width_____

Objective 2   Solve problems about quantities and their costs.

Video Examples

Review this example for Objective 2:

2.   The total receipts for a basketball game were
     $4690.50. There were 723 tickets sold, some for
     children and some for adults. If the adult tickets
     cost $9.50 and the children's tickets cost $4, how
     many of each type were there?

Step1 Read the problem.

Step 2 Assign variables. Let x = the number of
adult tickets and y = the number of children's
tickets.

Step 3 Write two equations. The total number of
tickets is 723. The total value of the tickets is
$4690.50.

$$x + y = 723 \qquad (1)$$
$$9.50x + 4y = 4690.5 \quad (2)$$

Step 4 Solve the system. Solve using
elimination. Multiply equation (1) by –4.

$$-4x - 4y = -2892$$
$$\underline{9.50x + 4y = 4690.5}$$
$$5.50x \qquad = 1798.50$$
$$x = 327$$

Substitute 327 for x in equation (1) to find y.

$$x + y = 723$$
$$327 + y = 723$$
$$y = 396$$

Now Try:

2.   Twice as many general
     admission tickets to a basketball
     game were sold as reserved seat
     tickets. General admission
     tickets cost $10 and reserved
     seat tickets cost $15. If the total
     value of both kinds of tickets
     was $26,250, how many tickets
     of each kind were sold?

     _____

Step 5 State the answer. The number of adult tickets is 327 and the number of children's tickets is 396.

Step 6 Check. The sum of the tickets is 723. The value of the tickets is

$$9.50(327) + 4(396) = 4690.50$$

This checks.

Objective 2 Practice Exercises

For extra help, see Example 2 on pages 325–326 of your text.

Write a system of equations for each problem, then solve the problem.

4.   There were 411 tickets sold for a soccer game, some for students and some for nonstudents. Student tickets cost $4.25 and nonstudent tickets cost $8.50 each. The total receipts were $3021.75. How many of each type were sold?

4.

student tix_____

nonstudent tix_____

5.   A cashier has some $5 bills and some $10 bills. The total value of the money is $750. If the number of tens is equal to twice the number of fives, how many of each type are there?

5.

$5 bills _____

$10 bills _____

6. Luke plans to buy 10 ties with exactly $162. If some   **6.**
    ties cost $14, and the others cost $25, how many ties
    of each price should he buy? $14 ties _____

                                                                 $25 ties _____

Objective 3   Solve problems about mixtures.

Video Examples

Review this example for Objective 3:

3. A mixture of 75% solution should be mixed with
a 55% solution to get 70 liters of 63% solution.
Determine the number of liters required of the
55% and 75% solutions.

Step 1 Read the problem carefully.

Step 2 Assign variables. Let $x =$ the number of
liters of the 75% liquid and $y =$ the number of
liters of the 55% liquid.

Step 3 Write two equations. The total amount of
liquid of the final mixture is 70 liters. The
amount of 75% solution mixed with 55%
solution will equal the 70 liters of 63% solution.

$$x + y = 70 \qquad\qquad (1)$$
$$0.75x + 0.55y = 0.63(70) \quad (2)$$

Step 4 Solve the system. Solve by substitution.
Solving equation (1) for x results in $70 - y$.

$$0.75x + 0.55y = 0.63(70)$$
$$0.75(70 - y) + 0.55y = 44.1$$
$$52.5 - 0.75y + 0.55y = 44.1$$
$$-0.20y + 52.5 = 44.1$$
$$-0.20y = -8.4$$
$$y = 42$$

Now Try:

3. A mixture of 85% solution
should be mixed with a 65%
solution to get 80 ounces of 77%
solution. Determine the number
of ounces required of the 85%
and 65% solutions.

_____

Substitute 42 for y in equation (1).
$$x = 70 - 42 = 28$$

Step 5 State the answer. There should be 28 liters of 75% solution and 42 liters of 55% solution.

Step 6 Check the answer in the original problems. $28 + 42 = 70$ and
$0.75(28) + 0.55(42) = 44.1$ The answer checks.

Objective 3 Practice Exercises

For extra help, see Example 3 on pages 326–327 of your text.

Write a system of equations for each problem, then solve the problem.

7.   Jorge wishes to make 150 pounds of coffee blend that can be sold for $8 per pound. The blend will be a mixture of coffee worth $6 per pound and coffee worth $12 per pound. How many pounds of each kind of coffee should be used in the mixture?

     7.

     $6 coffee_____

     $12 coffee_____

8.   A solution of 50% acid is mixed with a 10% acid solution to get a 500 mL solution that is 40% acid. Determine the number of milliliters required of the 50% and 10% solutions.

     8.

     50% solution _____

     10% solution _____

9. Ben wishes to blend candy selling for $1.60 a pound with candy selling for $2.50 a pound to get a mixture that will be sold for $1.90 a pound. How many pounds of the $1.60 and the $2.50 candy should be used to get 30 pounds of the mixture?

9.

$1.60 candy _____

$2.50 candy _____

Objective 4   Solve problems about distance, rate (or speed), and time.

Video Examples

Review this example for Objective 4:

4. Bill and Hillary start in Washington and fly in opposite directions. At the end of 4 hours, they are 4896 kilometers apart. If Bill flies 60 kilometers per hour faster than Hillary, what are their speeds?

Step 1 Read the problem carefully.

Step 2 Assign variables. Let x = Bill's rate of speed and y = Hillary's rate of speed.

Step 3 Write two equations.
$$x = 60 + y \qquad (1)$$
$$4x + 4y = 4896 \quad (2)$$

Step 4 Solve the system. . Solve by substitution.
$$4(60 + y) + 4y = 4896$$
$$240 + 4y + 4y = 4896$$
$$240 + 8y = 4896$$
$$8y = 4656$$
$$y = 582$$

Substitute 582 for y in equation (1).
$$x = 60 + 582 = 642$$

Step 5 State the answer. Bill's rate is 642 kmh and Hillary's rate is 582 kmh.

Step 6 Check. Since 4(642) + 4(582) = 4896 and $642 = 582 + 60$ the answers check.

Now Try:

4. Enid leaves Cherry Hill, driving by car toward New York, which is 90 miles away. At the same time, Jerry, riding his bicycle, leaves New York cycling toward Cherry Hill. Enid is traveling 28 miles per hour faster than Jerry. They pass each other $1\frac{1}{2}$ hours later. What are their speeds?

_____

Objective 4 Practice Exercises

For extra help, see Examples 4–5 on pages 328–329 of your text.

Write a system of equations for each problem, and then solve the problem.

10. It takes Carla's boat $\frac{1}{2}$ hour to go 8 miles
     downstream and 1 hour to make the return trip
     upstream. Find the speed of the current and the
     speed of Carla's boat in still water.

10.

boat speed_____

current speed _____

11. Two planes left Philadelphia traveling in opposite
     directions. Plane A left 15 minutes before plane B.
     After plane B had been flying for 1 hour, the planes
     were 860 miles apart. What were the speeds of the
     two planes if plane A was flying 40 miles per hour
     faster than plane B?

11.

plane A_____

plane B _____

12. At the beginning of a fund-raising walk, Steve and
     Vic are 30 miles apart. If they leave at the same time
     and walk in the same direction, Steve would
     overtake Vic in 15 hours. If they walked toward each
     other, they would meet in 3 hours. What are their
     speeds?

12.

Steve_____

Vic _____

Chapter 4 SYSTEMS OF LINEAR EQUATIONS AND INEQUALITIES

4.5   Solving Systems of Linear Inequalities

Learning Objectives
1   Solve systems of linear inequalities by graphing.

Key Terms

Use the vocabulary terms listed below to complete each statement in exercises 1−2.

> **system of linear inequalities**
>
> **solution set of a system of linear inequalities**

1.   All ordered pairs that make all inequalities of the system true at the same time is called the _____.

2.   A _____ contains two or more linear inequalities (and no other kinds of inequalities).

Objective 1   Solve systems of linear inequalities by graphing.

Video Examples

Review these examples for Objective 1:

1.   Graph the solution set of the system.

$$x + y \leq 3$$
$$5x - y \geq 5$$

To graph $x + y \leq 3$, graph the solid boundary line $x + y = 3$ using the intercepts (0, 3) and (3, 0). Determine the region to shade using (0, 0) as a test point.

$$x + y \leq 3$$
$$0 + 0 \overset{?}{\leq} 3$$
$$0 \leq 3 \quad \text{True}$$

Shade the region containing (0, 0).

To graph $5x - y \geq 5$, graph the solid boundary line $5x - y = 5$ using the intercepts (0, −5) and (1, 0). Determine the region to shade using (0, 0) as a test point.

$$5x - y \geq 5$$
$$5(0) - 0 \overset{?}{\geq} 5$$
$$0 \geq 5 \quad \text{False}$$

Shade the region that does not contain (0, 0).

Now Try:

1.   Graph the solution set of the system.

$$3x - y \leq 3$$
$$x + y \leq 0$$

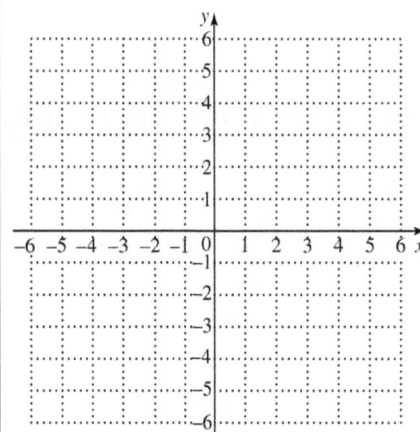

The solution set of this system includes all points in the intersection (overlap) of the graph of the two inequalities. This intersection is the gray shaded region and portions of the two boundary lines that surround it.

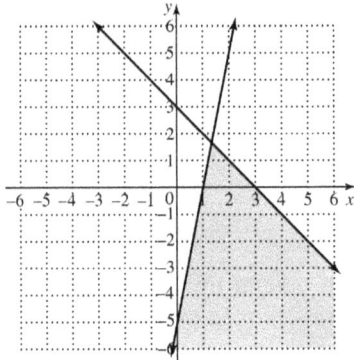

2. Graph the solution set of the system.

$$4x - y > 2$$

$$x + y > -2$$

Graph $4x - y > 2$ using a dashed line for $4x - y = 2$ and intercepts $(0, -2)$ and $\left(\frac{1}{2}, 0\right)$.

Using the test point $(0, 0)$, we shade the region that does not contain the point $(0, 0)$.

Graph $x + y > -2$ using a dashed line for $x + y = -2$ and intercepts $(-2, 0)$ and $(0, -2)$.
Using the test point $(0, 0)$, we shade the region that does contain the point $(0, 0)$.

The solution set is marked in gray. The solution set does not include either boundary line.

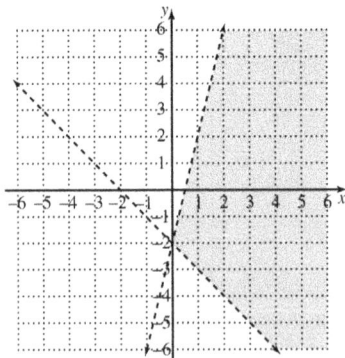

2. Graph the solution set of the system.

$$6x - y > 6$$

$$2x + 5y < 10$$

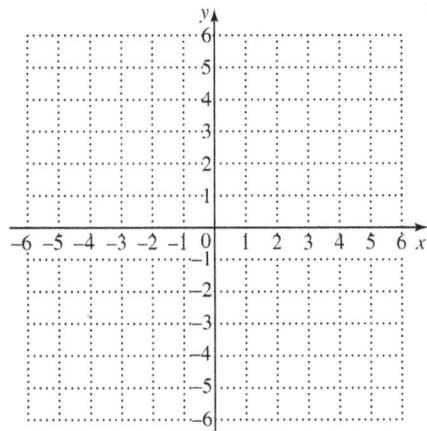

3. Graph the solution set of the system.

$$y \geq -1$$

$$2x - y > -1$$

Recall that $y = -1$ is a horizontal line through the point $(0, -1)$. Use a solid line, and shade the region with the point $(0, 0)$.

The graph of $2x - y > -1$ is created using a dashed line through the intercepts $(0, 1)$ and $\left(-\frac{1}{2}, 0\right)$. Shade the region with the point $(0, 0)$.

The solution set is marked in gray. The solution set includes the boundary line $y \geq -1$.

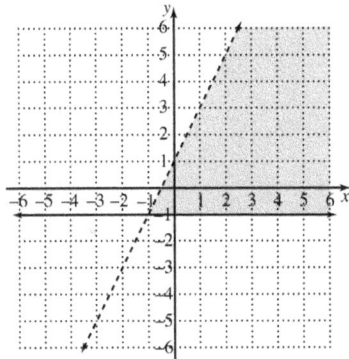

3. Graph the solution set of the system.

$$x - 3y \leq -7$$

$$x < 2$$

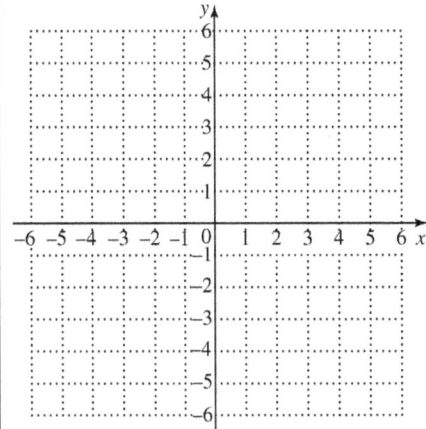

Objective 1 Practice Exercises

For extra help, see Examples 1–3 on pages 336–338 of your text.

Graph the solution of each system of linear inequalities.

1. $4x + 5y \leq 20$

$$y \leq x + 3$$

1.

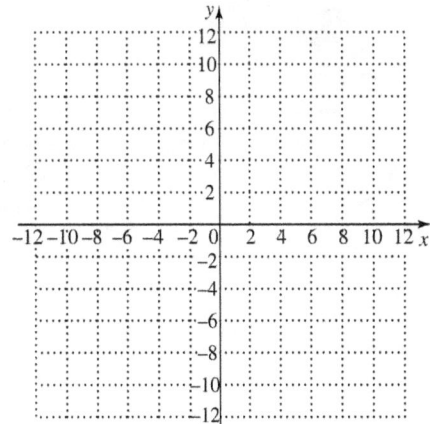

2. $x < 2y + 3$

$\quad\ 0 < x + y$

2.

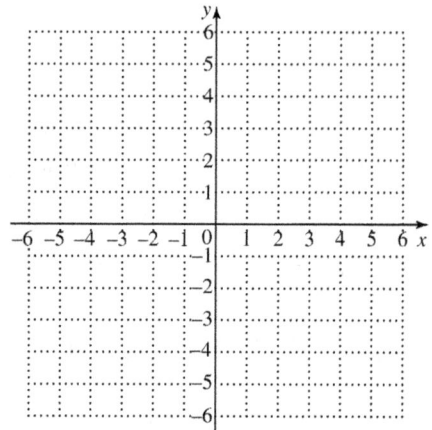

3. $y < 4$

$\quad\ x \geq -3$

3.

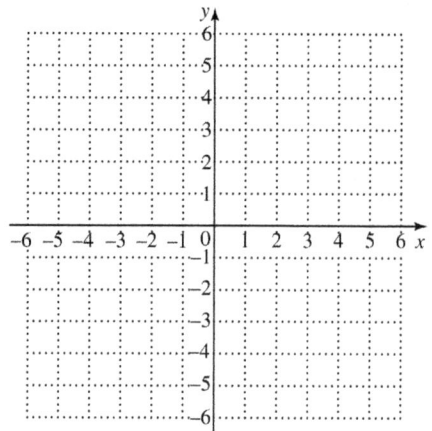

Chapter 5 EXPONENTS AND POLYNOMIALS

5.1   The Product Rule and Power Rules for Exponents

Learning Objectives
1   Use exponents.
2   Use the product rule for exponents.
3   Use the rule $\left(a^m\right)^n = a^{mn}$.
4   Use the rule $\left(ab\right)^m = a^m b^m$.
5   Use the rule $\left(\dfrac{a}{b}\right)^m = \dfrac{a^m}{b^m}$.
6   Use combinations of the rules for exponents.
7   Use the rules for exponents in a geometry problem.

Key Terms

Use the vocabulary terms listed below to complete each statement in exercises 1−3.

> **exponential expression**   **base**   **power**

1. 2^5 is read "2 to the fifth _____".

2.   A number written with an exponent is called a(n) _____.

3.   The _____ is the number being multiplied repeatedly.

Objective 1   Use exponents.

Video Examples

Review these examples for Objective 1:

1.   Write $5 \cdot 5 \cdot 5$ in exponential form and evaluate.

   Since 5 occurs as a factor three times, the base is 5 and the exponent is 3.

   $$5 \cdot 5 \cdot 5 = 5^3 = 125$$

2.   Name the base and exponent of each expression. Then evaluate.

   a. 3^4

   Base: 3
   Exponent: 4
   Value: $3^4 = 3 \cdot 3 \cdot 3 \cdot 3 = 81$

Now Try:

1.   Write $4 \cdot 4 \cdot 4 \cdot 4 \cdot 4$ in exponential form and evaluate.

   _____

2.   Name the base and exponent of each expression. Then evaluate.

   a. 2^6

   _____

b. -3^4

Base: 3
Exponent: 4
Value: $-3^4 = -1 \cdot (3 \cdot 3 \cdot 3 \cdot 3) = -81$

c. $(-3)^4$

Base: −3
Exponent: 4
Value: $(-3)^4 = (-3)(-3)(-3)(-3) = 81$

b. -2^6

_____

c. $(-2)^6$

_____

Objective 1 Practice Exercises

For extra help, see Examples 1–2 on page 352 of your text.

Write the expression in exponential form.

1. $\left(\dfrac{1}{3}\right)\left(\dfrac{1}{3}\right)\left(\dfrac{1}{3}\right)\left(\dfrac{1}{3}\right)\left(\dfrac{1}{3}\right)$

1. _____

Evaluate each exponential expression. Name the base and the exponent.

2. $(-4)^4$

2. _____

base_____

exponent_____

3. -3^8

3. _____

base_____

exponent_____

Objective 2   Use the product rule for exponents.

Video Examples

Review these examples for Objective 2:

3. Use the product rule for exponents to simplify, if possible.

    a. $8^4 \cdot 8^5$

        $8^4 \cdot 8^5 = 8^{4+5}$

              $= 8^9$

Now Try:

3. Use the product rule for exponents to simplify, if possible.

    a. $9^6 \cdot 9^7$

_____

d. $m^7 m^8 m^9$

$$m^7 m^8 m^9 = m^{7+8+9}$$
$$= m^{24}$$

d. $m^{11} m^9 m^7$

_____

4. Multiply $5x^4$ and $6x^9$.

$$5x^4 \cdot 6x^9 = (5 \cdot 6) \cdot (x^4 \cdot x^9)$$
$$= 30x^{4+9}$$
$$= 30x^{13}$$

4. Multiply $6x^5$ and $3x^6$.

_____

3f. Use the product rule for exponents to simplify, if possible.

$$5^2 + 5^3$$

$$5^2 + 5^3 = 25 + 125$$
$$= 150$$

3f. Use the product rule for exponents to simplify, if possible.

$$3^4 + 3^3$$

_____

Objective 2 Practice Exercises

For extra help, see Examples 3–4 on pages 353–354 of your text.

Use the product rule to simplify each expression, if possible. Write each answer in exponential form.

4. $7^4 \cdot 7^3$

4. _____

5. $\left(-2c^7\right)\left(-4c^8\right)$

5. _____

6. $\left(3k^7\right)\left(-8k^2\right)\left(-2k^9\right)$

6. _____

Objective 3   Use the rule $\left(a^m\right)^n = a^{mn}$.

Video Examples

Review these examples for Objective 3:

5.   Use power rule (a) for exponents to simplify.

    a. $\left(3^4\right)^5$

$$\left(3^4\right)^5 = 3^{4\cdot5}$$
$$= 3^{20}$$

    c. $\left(x^3\right)^4$

$$\left(x^3\right)^4 = x^{3\cdot4}$$
$$= x^{12}$$

Now Try:

5.   Use power rule (a) for exponents to simplify.

    a. $\left(4^5\right)^3$

    _____

    c. $\left(x^5\right)^6$

    _____

Objective 3 Practice Exercises

For extra help, see Example 5 on page 354 of your text.

Simplify each expression. Write all answers in exponential form.

7. $\left(7^3\right)^4$

    7.   _____

8. $-\left(v^4\right)^9$

    8.   _____

9. $\left[(-3)^3\right]^7$

    9.   _____

Objective 4   Use the rule $\left(ab\right)^m = a^m b^m$.

Video Examples

Review this example for Objective 4:

6a. Use power rule (b) for exponents to simplify.

$$\left(5xy\right)^3$$

$$\left(5xy\right)^3 = 5^3 x^3 y^3$$
$$= 125 x^3 y^3$$

Now Try:

6a. Use power rule (b) for exponents to simplify.

$$\left(4ab\right)^3$$

    _____

Objective 4 Practice Exercises

For extra help, see Example 6 on page 355 of your text.

Simplify each expression.

10. $\left(5r^3t^2\right)^4$ 10. _____

11. $\left(-0.2a^4b\right)^3$ 11. _____

12. $\left(-2w^3z^7\right)^4$ 12. _____

Objective 5   Use the rule $\left(\dfrac{a}{b}\right)^m = \dfrac{a^m}{b^m}$.

Video Examples

Review this example for Objective 5:   **Now Try:**
7b. Use power rule (c) for exponents to simplify.   **7b.** Use power rule (c) for exponents to simplify.

$$\left(\frac{1}{8}\right)^3$$

$$\left(\frac{1}{4}\right)^5$$

$$\left(\frac{1}{8}\right)^3 = \frac{1^3}{8^3} = \frac{1}{512}$$

_____

Objective 5 Practice Exercises

For extra help, see Example 7 on page 356 of your text.

Simplify each expression.

13. $\left(-\dfrac{2x}{5}\right)^3$ 13. _____

14. $\left(\dfrac{xy}{z^2}\right)^4$ 14. _____

15. $\left(\dfrac{-2a}{b^2}\right)^7$ 15. _____

Name: Date:
Instructor: Section:

Objective 6   Use combinations of the rules for exponents.

Video Examples

Review these examples for Objective 6:

8. Simplify each expression.

a. $\left(\dfrac{3}{4}\right)^3 \cdot 3^2$

$$\left(\dfrac{3}{4}\right)^3 \cdot 3^2 = \dfrac{3^3}{4^3} \cdot \dfrac{3^2}{1}$$

$$= \dfrac{3^3 \cdot 3^2}{4^3 \cdot 1}$$

$$= \dfrac{3^{3+2}}{4^3}$$

$$= \dfrac{3^5}{4^3}, \quad \text{or} \quad \dfrac{243}{64}$$

d. $\left(-x^5 y\right)^4 \left(-x^6 y^5\right)^3$

$$\left(-x^5 y\right)^4 \left(-x^6 y^5\right)^3$$

$$= \left(-1x^5 y\right)^4 \left(-1x^6 y^5\right)^3$$

$$= (-1)^4 \left(x^5\right)^4 \left(y^4\right) \cdot (-1)^3 \left(x^6\right)^3 \left(y^5\right)^3$$

$$= (-1)^4 \left(x^{20}\right)\left(y^4\right) \cdot (-1)^3 \left(x^{18}\right)\left(y^{15}\right)$$

$$= (-1)^7 x^{20+18} y^{4+15}$$

$$= -1x^{38} y^{19}$$

$$= -x^{38} y^{19}$$

Now Try:

8. Simplify each expression.

a. $\left(\dfrac{5}{2}\right)^3 \cdot 5^2$

_____

d. $\left(-x^5 y\right)^3 \left(-x^6 y^5\right)^2$

_____

Objective 6 Practice Exercises

For extra help, see Example 8 on pages 356–357 of your text.

Simplify. Write all answers in exponential form.

16. $\left(-x^3\right)^2 \left(-x^5\right)^4$

16. _____

17. $\left(2ab^2c\right)^5(ab)^4$

17. _____

18. $\left(5x^2y^3\right)^7\left(5xy^4\right)^4$

18. _____

Objective 7   Use the rules for exponents in a geometry problem.

Video Examples

Review this example for Objective 7:
9a. Find the area of the figure.

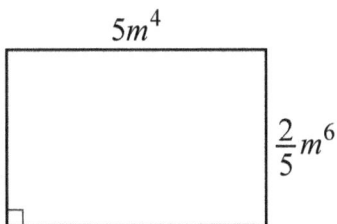

$5m^4$

$\frac{2}{5}m^6$

Use the formula for the area of a rectangle.
$A = LW$

$A = \left(5m^4\right)\left(\frac{2}{5}m^6\right)$

$A = 5\cdot\frac{2}{5}\cdot m^{4+6}$

$A = 2m^{10}$

Now Try:
9a. Find the area of the figure.

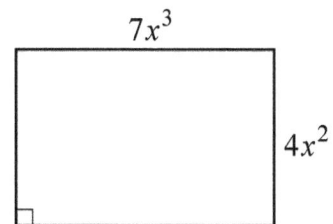

$7x^3$

$4x^2$

_____

Objective 7 Practice Exercises

For extra help, see Example 9 on page 357 of your text.

Find a polynomial that represents the area of each figure.

19.

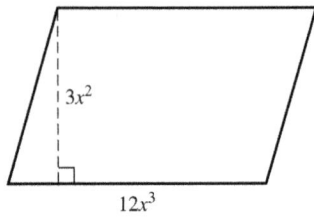

19. _____

20.

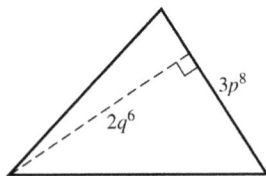

20. _____

21.

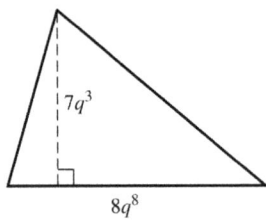

21. _____

Chapter 5 EXPONENTS AND POLYNOMIALS

5.2   Integer Exponents and the Quotient Rule

Learning Objectives
1   Use 0 as an exponent.
2   Use negative numbers as exponents.
3   Use the quotient rule for exponents.
4   Use combinations of the rules for exponents.

Key Terms

Use the vocabulary terms listed below to complete each statement in exercises 1−3.

exponent   **base**   **product rule for exponents**

power rule for exponents

1.   The statement "If m and n are any integers, then $\left(a^{m}\right)^{n} = a^{mn}$ " is an example of the

      _____.

2.   In the expression a^{m}, a is the _____ and m is the _____.

3.   The statement "If m and n are any integers, then $a^{m} \cdot a^{n} = a^{m+n}$ is an example of the

      _____.

Objective 1   Use 0 as an exponent.

Video Examples

Review these examples for Objective 1:
1.   Evaluate.

    a. $75^{0} = 1$

    b. $(-75)^{0} = 1$

    c. $-75^{0} = -(1)$ or -1

    d. $x^{0} = 1$ $(x \neq 0)$

Now Try:
1.   Evaluate.

    a. 88^{0}

    _____

    b. $(-88)^{0}$

    _____

    c. -88^{0}

    _____

    d. a^{0} $(a \neq 0)$

    _____

e. $9x^0 = 9(1)$, or 9 $(x \neq 0)$

e. $88a^0$ $(a \neq 0)$

_____

f. $(9x)^0 = 1$ $(x \neq 0)$

f. $(88a)^0$ $(a \neq 0)$

_____

Objective 1 Practice Exercises

For extra help, see Example 1 on page 360 of your text.

Evaluate each expression.

1. -12^0

1. _____

2. $-15^0 - (-15)^0$

2. _____

3. $\dfrac{0^8}{8^0}$

3. _____

Objective 2   Use negative numbers as exponents.

Video Examples

Review these examples for Objective 2:

2. Simplify by writing with positive exponents. Assume that all variables represent nonzero real numbers.

a. 5^{-2}

$5^{-2} = \dfrac{1}{5^2}$, or $\dfrac{1}{25}$

d. $\left(\dfrac{3}{5}\right)^{-4}$

$\left(\dfrac{3}{5}\right)^{-4} = \left(\dfrac{5}{3}\right)^4$ The reciprocal of $\dfrac{3}{5}$ is $\dfrac{5}{3}$.

$= \dfrac{5^4}{3^4}$

$= \dfrac{625}{81}$

Now Try:

2. Simplify by writing with positive exponents. Assume that all variables represent nonzero real numbers.

a. 8^{-2}

_____

d. $\left(\dfrac{6}{7}\right)^{-2}$

_____

f. $5^{-1} - 3^{-1}$

$$5^{-1} - 3^{-1} = \frac{1}{5} - \frac{1}{3}$$

$$= \frac{3}{15} - \frac{5}{15}$$

$$= -\frac{2}{15}$$

h. $\dfrac{1}{x^{-6}}$

$$\frac{1}{x^{-6}} = \frac{1^{-6}}{x^{-6}}$$

$$= \left(\frac{1}{x}\right)^{-6}$$

$$= x^6$$

3. Simplify. Assume that all variables represent nonzero real numbers.

b. $\dfrac{a^{-6}}{b^{-1}} = \dfrac{b^1}{a^6}$, or $\dfrac{b}{a^6}$

c. $\dfrac{x^{-3}y}{4z^{-4}} = \dfrac{yz^4}{4x^3}$

d. $\left(\dfrac{p}{3q}\right)^{-3}$

$$\left(\frac{p}{3q}\right)^{-3} = \left(\frac{3q}{p}\right)^3$$

$$= \frac{3^3 q^3}{p^3}$$

$$= \frac{27q^3}{p^3}$$

f. $4^{-1} - 8^{-1}$

_____

h. $\dfrac{1}{x^{-8}}$

_____

3. Simplify. Assume that all variables represent nonzero real numbers.

b. $\dfrac{x^{-7}}{y^{-1}}$

_____

c. $\dfrac{p^{-3}q}{4r^{-5}}$

_____

d. $\left(\dfrac{a}{5b}\right)^{-4}$

_____

Objective 2 Practice Exercises

For extra help, see Examples 2–3 on pages 361–363 of your text.

Evaluate or simplify each expression, and write it using only positive exponents. Assume that all variables represent nonzero real numbers.

4. $-2k^{-4}$

4. _____

5. $(m^2n)^{-9}$

5. _____

6. $\dfrac{2x^{-4}}{3y^{-7}}$

6. _____

Objective 3   Use the quotient rule for exponents.

Video Examples

Review these examples for Objective 3:

4.   Simplify. Assume that all variables represent nonzero real numbers.

   b. $\dfrac{5^3}{5^7} = 5^{3-7} = 5^{-4} = \dfrac{1}{5^4} = \dfrac{1}{625}$

   d. $\dfrac{p^6}{p^{-4}} = p^{6-(-4)} = p^{10}$

Now Try:

4.   Simplify. Assume that all variables represent nonzero real numbers.

   b. $\dfrac{2^6}{2^{10}}$

   _____

   d. $\dfrac{z^6}{z^{-4}}$

   _____

e. $\dfrac{4^3 x^6}{4^5 x^2}$

$$\dfrac{4^3 x^6}{4^5 x^2} = \dfrac{4^3}{4^5} \cdot \dfrac{x^6}{x^2}$$

$$= 4^{3-5} \cdot x^{6-2}$$

$$= 4^{-2} x^4$$

$$= \dfrac{x^4}{4^2}, \quad \text{or} \quad \dfrac{x^4}{16}$$

e. $\dfrac{5^4 a^6}{5^6 a^3}$

_____

Objective 3 Practice Exercises

For extra help, see Example 4 on page 364 of your text.

Use the quotient rule to simplify each expression, and write it using only positive exponents. Assume that all variables represent nonzero real numbers.

7. $\dfrac{4k^7 m^{10}}{8k^3 m^5}$

7. _____

8. $\dfrac{a^4 b^3}{a^{-2} b^{-3}}$

8. _____

9. $\dfrac{3^{-1} m^{-4} p^6}{3^4 m^{-1} p^{-2}}$

9. _____

Name: Date:
Instructor: Section:

Objective 4   Use combinations of the rules for exponents.

Video Examples

Review these examples for Objective 4:

5. Simplify each expression. Assume that all variables represent nonzero real numbers.

d. $\left(\dfrac{3x^4}{4}\right)^{-5}$

$\left(\dfrac{3x^4}{4}\right)^{-5} = \left(\dfrac{4}{3x^4}\right)^{5}$

$\phantom{\left(\dfrac{3x^4}{4}\right)^{-5}} = \dfrac{4^5}{3^5 x^{20}}$

$\phantom{\left(\dfrac{3x^4}{4}\right)^{-5}} = \dfrac{1024}{243 x^{20}}$

e. $\left(\dfrac{5x^{-3}}{2^{-1}y^4}\right)^{-2}$

$\left(\dfrac{5x^{-3}}{2^{-1}y^4}\right)^{-2} = \dfrac{5^{-2}x^6}{2^2 y^{-8}}$

$\phantom{\left(\dfrac{5x^{-3}}{2^{-1}y^4}\right)^{-2}} = \dfrac{x^6 y^8}{5^2 \cdot 2^2}$

$\phantom{\left(\dfrac{5x^{-3}}{2^{-1}y^4}\right)^{-2}} = \dfrac{x^6 y^8}{100}$

Now Try:

5. Simplify each expression. Assume that all variables represent nonzero real numbers.

d. $\left(\dfrac{2p^4}{3}\right)^{-5}$

_____

e. $\left(\dfrac{4x^{-4}}{5^{-1}y^5}\right)^{-3}$

_____

Objective 4 Practice Exercises

For extra help, see Example 5 on pages 365–366 of your text.

Simplify each expression, and write it using only positive exponents. Assume that all variables represent nonzero real numbers.

10. $(9xy)^7 (9xy)^{-8}$

10. _____

Name: Date:

Instructor: Section:

11. $\dfrac{\left(a^{-1}b^{-2}\right)^{-4}\left(ab^2\right)^6}{\left(a^3b\right)^{-2}}$

                                                            11. _____

12. $\left(\dfrac{k^3t^4}{k^2t^{-1}}\right)^{-4}$

                                                            12. _____

Chapter 5 EXPONENTS AND POLYNOMIALS

5.3   An Application of Exponents: Scientific Notation

Learning Objectives
1   Express numbers in scientific notation.
2   Convert numbers in scientific notation to standard notation.
3   Use scientific notation in calculations.

Key Terms

Use the vocabulary terms listed below to complete each statement in exercises 1–3.

scientific notation   quotient rule   power rule

1.   A number written as $a \times 10^n$, where $1 \le |a| < 10$ and n is an integer, is written in

_____.

2.   The statement "If m and n are any integers and $b \ne 0$, then $\left(\dfrac{a}{b}\right)^m = \dfrac{a^m}{b^m}$ " is an

example of the _____.

3.   The statement "If m and n are any integers and $b \ne 0$, then $\dfrac{a^m}{a^n} = a^{m-n}$ " is an

example of the _____.

Objective 1   Express numbers in scientific notation.

Video Examples

Review these examples for Objective 1:

1.   Write each number in scientific notation.

   b. 84,300,000,000

   Move the decimal point 10 places to the left.
      $84,300,000,000 = 8.43 \times 10^{10}$

   c. 0.00573

   The first nonzero digit is 5. Count the places.
   Move the decimal point 3 places to the right.
      $0.00573 = 5.73 \times 10^{-3}$

Now Try:

1.   Write each number in scientific notation.

   b. 47,710,000,000

      _____

   c. 0.0463

      _____

Objective 1 Practice Exercises

For extra help, see Example 1 on page 372 of your text.

Write each number in scientific notation.

1. 23,651

1. _____

2. −429,600,000,000

2. _____

3. −0.0002208

3. _____

Objective 2   Convert numbers in scientific notation to standard notation.

Video Examples

Review these examples for Objective 2:
2. Write each number in standard notation.

b. 3.57×10^6

Move the decimal point 6 places to the right, and add four zeros.

$$3.57 \times 10^6 = 3,570,000$$

d. -8.98×10^{-3}

Move the decimal point 3 places to the left.
$$-8.98 \times 10^{-3} = -0.00898$$

Now Try:
2. Write each number in standard notation.

b. 2.796×10^7

_____

d. -1.64×10^{-4}

_____

Objective 2 Practice Exercises

For extra help, see Example 2 on page 373 of your text.

Write each number in standard notation.

4. -2.45×10^6

4. _____

5. 6.4×10^{-3}

5. _____

6. -4.02×10^0

6. _____

Objective 3   Use scientific notation in calculations.

Video Examples

Review these examples for Objective 3:

3. Perform each calculation. Write answers in scientific notation and standard notation.

a. $(8 \times 10^4)(7 \times 10^3)$

$$(8 \times 10^4)(7 \times 10^3) = (8 \times 7)(10^4 \times 10^3)$$
$$= 56 \times 10^7$$
$$= (5.6 \times 10^1) \times 10^7$$
$$= 5.6 \times 10^8$$
$$= 560,000,000$$

b. $\dfrac{6 \times 10^{-4}}{3 \times 10^2}$

$$\frac{6 \times 10^{-4}}{3 \times 10^2} = \frac{6}{3} \times \frac{10^{-4}}{10^2}$$
$$= 2 \times 10^{-6}$$
$$= 0.000002$$

4. A light-year is the distance that light travels in one year. The speed of light is about 3×10^5 km per second. How many kilometers are in a light-year?

Convert km per second to km per year.

$$\frac{3 \times 10^5 \text{km}}{1 \text{ sec}} \cdot \frac{60 \text{ sec}}{1 \text{ min}} \cdot \frac{60 \text{ min}}{1 \text{ hr}} \cdot \frac{24 \text{ hr}}{1 \text{ day}} \cdot \frac{365 \text{ day}}{1 \text{ yr}}$$

$$= (3 \times 60^2 \times 24 \times 365) \times 10^5$$
$$= 94608000 \times 10^5$$
$$= (9.4608 \times 10^7) \times 10^5$$
$$\approx 9.46 \times 10^{12} \text{ km/yr}$$

Thus, there are 9.46×10^{12} km in a light-year.

Now Try:

3. Perform each calculation. Write answers in scientific notation and standard notation.

a. $(9 \times 10^5)(3 \times 10^2)$

_____

b. $\dfrac{39 \times 10^{-3}}{13 \times 10^5}$

_____

4. Earth has a mass of 6×10^{24} kilograms and a volume of 1.1×10^{21} cubic meters. What is Earth's density in kilograms per cubic meter? Round to the nearest hundredth.

_____

Name: Date:
Instructor: Section:

Objective 3 Practice Exercises

For extra help, see Examples 3–5 on pages 373–374 of your text.

Perform the indicated operations, and write the answers in scientific notation.

7. $(2.3 \times 10^4) \times (1.1 \times 10^{-2})$

7. _____

8. $\dfrac{9.39 \times 10^1}{3 \times 10^3}$

8. _____

Work the problem. Give answer in scientific notation.

9.   There are about 6×10^{23} atoms in a mole of atoms. How many atoms are there in 8.1×10^{-5} mole?

9. _____

Chapter 5 EXPONENTS AND POLYNOMIALS

5.4   Adding and Subtracting Polynomials

Learning Objectives
1   Identify terms and coefficients.
2   Combine like terms.
3   Know the vocabulary for polynomials.
4   Evaluate polynomials.
5   Add polynomials.
6   Subtract polynomials.
7   Add and subtract polynomials with more than one variable.

Key Terms

Use the vocabulary terms listed below to complete each statement in exercises 1−9.

term	**like terms**	**polynomial**
descending powers		**degree of a term**
degree of a polynomial		**monomial**
binomial		**trinomial**

1.   The _____ is the sum of the exponents on the variables in that term.

2.   A polynomial in x is written in _____ if the exponents on x in its terms are decreasing order.

3.   A _____ is a number, a variable, or a product or quotient of a number and one or more variables raised to powers.

4.   A polynomial with exactly three terms is called a _____.

5.   A _____ is a term, or the sum of a finite number of terms with whole number exponents.

6.   A polynomial with exactly one term is called a _____.

7.   The _____ is the greatest degree of any term of the polynomial.

8.   A _____ is a polynomial with exactly two terms.

9.   Terms with exactly the same variables (including the same exponents) are called _____.

Name: _____ Date: _____

Instructor: _____ Section: _____

Objective 1   Identify terms and coefficients.

Video Examples

Review these examples for Objective 1:

1. Identify the coefficient of each term in the expression. Then give the number of terms.

   a. $-4x^3 + x - 7$

   $-4x^3 + x - 7 = -4x^3 + 1x + (-7x^0)$
   The coefficients are –4, 1, and –7.
   There are 3 terms: $-4x^3$, x, and -7.

   b. $6 - y^2$

   $6 - y^2 = 6y^0 + (-1)y^2$
   The coefficients are 6 and –1.
   There are 2 terms.

Now Try:

1. Identify the coefficient of each term in the expression. Then give the number of terms.

   a. $-2w^5 - w + 3$

   _____

   b. $-8 + x^4$

   _____

Objective 1 Practice Exercises

For extra help, see Example 1 on page 379 of your text.

Identify the coefficient of each term in the expression. Then give the number of terms.

1. x^6

   1. _____

2. $\dfrac{m}{7}$

   2. _____

3. $\dfrac{2}{5}y^2 - 3y^3 + y^4$

   3. _____

Name: Date:
Instructor: Section:

Objective 2   Combine like terms.

Video Examples

Review these examples for Objective 2:	Now Try:
2. Simplify each expression by combining like terms.	2. Simplify each expression by combining like terms.

Review these examples for Objective 2:

2.   Simplify each expression by combining like terms.

   e. $19m^3 + 6m + 5m^3$

   $19m^3 + 6m + 5m^3 = (19+5)m^3 + 6m$

   $= 24m^3 + 6m$

   b. $4x^5 - 15x^5 + x^5$

   $4x^5 - 15x^5 + x^5 = (4-15+1)x^5$

   $= -10x^5$

Now Try:

2.   Simplify each expression by combining like terms.

   e. $22m^2 + 15m^3 + 7m^2$

   _____

   b. $9x^7 - 18x^7 + x^7$

   _____

Objective 2 Practice Exercises

For extra help, see Example 2 on page 380 of your text.

In each polynomial, combine like terms whenever possible. Write the result with descending powers.

4. $7z^3 - 4z^3 + 5z^3 - 11z^3$ 4.   _____

5. $-1.3z^7 + 0.4z^7 + 2.6z^8$ 5.   _____

6. $6c^3 - 9c^2 - 2c^2 + 14 + 3c^2 - 6c - 8 + 2c^3$ 6.   _____

Objective 3   Know the vocabulary for polynomials.

Video Examples

Review these examples for Objective 3:	**Now Try:**
3. Simplify each polynomial if possible. Then give the degree and tell whether the polynomial is a monomial, a binomial, a trinomial, or none of these.	3. Simplify each polynomial if possible. Then give the degree and tell whether the polynomial is a monomial, a binomial, a trinomial, or none of these.
a. $5x^4 + 7x$	**a.** $8x^3 + 4x^2 + 6$
We cannot simplify further. This is a binomial of degree 4.	_____
d. $9x - 7x + 3x$	**d.** $x^5 + 3x^5$
$9x - 7x + 3x = 5x$ The degree is 1. The simplified polynomial is a monomial.	_____

Objective 3 Practice Exercises

For extra help, see Example 3 on page 381 of your text.

For each polynomial, first simplify, if possible, and write the resulting polynomial in descending powers of the variable. Then give the degree of this polynomial, and tell whether it is a monomial, a binomial, a trinomial, or none of these.

7. $3n^8 - n^2 - 2n^8$

                                                       7.   _____

                                                       degree: _____

                                                       type: _____

8. $-d^2 + 3.2d^3 - 5.7d^8 - 1.1d^5$

                                                       8.   _____

                                                       degree: _____

                                                       type: _____

9. $-6c^4 - 6c^2 + 9c^4 - 4c^2 + 5c^5$

                                                       9.   _____

                                                       degree: _____

                                                       type: _____

Objective 4   Evaluate polynomials.

Video Examples

Review these examples for Objective 4:

4.   Find the value of $4x^3 + 6x^2 - 5x - 5$ for

   a. $x = -3$

$$4x^3 + 6x^2 - 5x - 5$$
$$= 4(-3)^3 + 6(-3)^2 - 5(-3) - 5$$
$$= 4(-27) + 6(9) - 5(-3) - 5$$
$$= -108 + 54 + 15 - 5$$
$$= -44$$

   b. $x = 2$

$$4x^3 + 6x^2 - 5x - 5$$
$$= 4(2)^3 + 6(2)^2 - 5(2) - 5$$
$$= 4(8) + 6(4) - 5(2) - 5$$
$$= 32 + 24 - 10 - 5$$
$$= 41$$

Now Try:

4.   Find the value of

$5x^4 + 3x^2 - 9x - 7$ for

   a. $x = 4$

   _____

   b. $x = -4$

   _____

Objective 4 Practice Exercises

For extra help, see Example 4 on page 382 of your text.

Find the value of each polynomial (a) *when x = –2 and* (b) *when x = 3.*

10. $3x^3 + 4x - 19$

11. $-4x^3 + 10x^2 - 1$

10. a. _____

     b. _____

11. a. _____

     b. _____

12. $x^4 - 3x^2 - 8x + 9$ **12. a.**_____

                                                b._____

Objective 5   Add polynomials.

Video Examples

Review these examples for Objective 5:

5a. Add vertically.

$5x^4 - 7x^3 + 9$ and $-3x^4 + 8x^3 - 7$

Write like terms in columns.

$5x^4 - 7x^3 + 9$
$\underline{-3x^4 + 8x^3 - 7}$

Now add, column by column.

$$\begin{array}{ccc} 5x^4 & -7x^3 & 9 \\ \underline{-3x^4} & \underline{8x^3} & \underline{-7} \\ 2x^4 & x^3 & 2 \end{array}$$

Add the three sums together to obtain the answer.

$2x^4 + x^3 + 2$

6b. Find the sum.

$\left(5x^4 - 7x^2 + 6x\right) + \left(-3x^3 + 4x^2 - 7\right)$

$\left(5x^4 - 7x^2 + 6x\right) + \left(-3x^3 + 4x^2 - 7\right)$

$= 5x^4 - 3x^3 - 7x^2 + 4x^2 + 6x - 7$

$= 5x^4 - 3x^3 - 3x^2 + 6x - 7$

Now Try:

5a. Add vertically.

$8x^3 - 9x^2 + x$ and
$-3x^3 + 4x^2 + 3x$

_____

6b. Find the sum.

$\left(8x^2 - 6x + 4\right) + \left(7x^3 - 8x - 5\right)$

_____

Name: _____   Date: _____
Instructor: _____   Section: _____

Objective 5 Practice Exercises

For extra help, see Examples 5–6 on pages 382–383 of your text.

Add.

13. $9m^3 + 4m^2 - 2m + 3$ 13. _____
        $\underline{-4m^3 - 6m^2 - 2m + 1}$

14. $\left(x^2 + 6x - 8\right) + \left(3x^2 - 10\right)$ 14. _____

15. $\left(3r^3 + 5r^2 - 6\right) + \left(2r^2 - 5r + 4\right)$ 15. _____

Objective 6   Subtract polynomials.

Video Examples

Review this example for Objective 6:
7b. Perform the subtraction.

Subtract $8x^3 - 5x^2 + 8$ from $9x^3 + 6x^2 - 7$.

$\left(9x^3 + 6x^2 - 7\right) - \left(8x^3 - 5x^2 + 8\right)$
$\quad = \left(9x^3 + 6x^2 - 7\right) + \left(-8x^3 + 5x^2 - 8\right)$
$\quad = x^3 + 11x^2 - 15$

Now Try:
7b. Perform the subtraction.

$\left(7x^3 - 3x - 5\right) - \left(18x^3 + 4x - 6\right)$

_____

Name: _____   Date: _____

Instructor: _____   Section: _____

Objective 6 Practice Exercises

For extra help, see Examples 7–9 on pages 383–384 of your text.

Subtract.

16. $\left(-8w^3 + 11w^2 - 12\right) - \left(-10w^2 + 3\right)$ 16. _____

17. $\left(8b^4 - 4b^3 + 7\right) - \left(2b^2 + b + 9\right)$ 17. _____

18. $\left(9x^3 + 7x^2 - 6x + 3\right) - \left(6x^3 - 6x + 1\right)$ 18. _____

Objective 7   Add and subtract polynomials with more than one variable.

Video Examples

Review this example for Objective 7:
10b.   Add or subtract as indicated.

$$\left(3x^2y + 5xy + y^2\right) - \left(4x^2y + xy - 3y^2\right)$$

$$\left(3x^2y + 5xy + y^2\right) - \left(4x^2y + xy - 3y^2\right)$$
$$= 3x^2y + 5xy + y^2 - 4x^2y - xy + 3y^2$$
$$= -x^2y + 4xy + 4y^2$$

Now Try:
10b.   Add or subtract as indicated.

$$\left(7x^2y + 3xy + 4y^2\right)$$
$$-\left(6x^2y - xy + 4y^2\right)$$

_____

Name: Date:

Instructor: Section:

Objective 7 Practice Exercises

For extra help, see Example 10 on page 384 of your text.

Add or subtract as indicated.

19. $\left(-2a^6 + 8a^4b - b^2\right) - \left(a^6 + 7a^4b + 2b^2\right)$ 19. _____

20. $\left(4ab + 2bc - 9ac\right) + \left(3ca - 2cb - 9ba\right)$ 20. _____

21. $\left(2x^2y + 2xy - 4xy^2\right) + \left(6xy + 9xy^2\right) - \left(9x^2y + 5xy\right)$ 21. _____

Chapter 5 EXPONENTS AND POLYNOMIALS

5.5   Multiplying Polynomials

Learning Objectives
1   Multiply monomials.
2   Multiply a monomial and a polynomial.
3   Multiply two polynomials.
4   Multiply binomials using the FOIL method.

Key Terms

Use the vocabulary terms listed below to complete each statement in exercises 1−3.

FOIL   outer product   inner product

1.   The _____ of $(2y-5)(y+8)$ is $-5y$.

2.   _____ is a shortcut method for finding the product of two binomials.

3.   The _____ of $(2y-5)(y+8)$ is $16y$.

Objective 1   Multiply monomials.

Video Examples

Review this example for Objective 1:
1a. Find the product.

$$3w^2(-7w)$$

Use the commutative and associative properties.
$$3w^2(-7w) = 3(-7)\cdot w^2 \cdot w$$
$$= -21w^{2+1}$$
$$= -21w^3$$

Now Try:
1a. Find the product.

$$9p^3(-8p)$$

_____

Objective 1 Practice Exercises

For extra help, see Example 1 on page 389 of your text.

Find each product.

1. $-5m^2(-2m^2)$

1.   _____

2. $4x^4\left(6x^2\right)$

2. _____

3. $-3ab^2\left(9a^3b\right)$

3. _____

Objective 2   Multiply a monomial and a polynomial.

Video Examples

Review this example for Objective 2:

2a. Find each product.

$5x^2\left(7x+3\right)$

Use the distributive property.
$$5x^2\left(7x+3\right)=5x^2\left(7x\right)+5x^2\left(3\right)$$
$$=35x^3+15x^2$$

Now Try:

2a. Find each product.

$8x^3\left(4x+8\right)$

_____

Objective 2 Practice Exercises

For extra help, see Example 2 on page 389 of your text.

Find each product.

4. $7z\left(5z^3+2\right)$

4. _____

5. $2m\left(3+7m^2+3m^3\right)$

5. _____

6. $-3y^2\left(2y^3+3y^2-4y+11\right)$

6. _____

Objective 3 Multiply two polynomials.

Video Examples

Review these examples for Objective 3:

Now Try:

3. Multiply $(x^2 + 6)(5x^3 - 4x^2 + 3x)$.

Multiply each term of the second polynomial by each term of the first.

$(x^2 + 6)(5x^3 - 4x^2 + 3x)$

$= x^2(5x^3) + x^2(-4x^2) + x^2(3x)$

$\qquad + 6(5x^3) + 6(-4x^2) + 6(3x)$

$= 5x^5 - 4x^4 + 3x^3 + 30x^3 - 24x^2 + 18x$

$= 5x^5 - 4x^4 + 33x^3 - 24x^2 + 18x$

3. Multiply
$(x^3 + 9)(4x^4 - 2x^2 + x)$

_____

4. Multiply $(2x^3 + 7x^2 + 5x - 1)(4x + 6)$ vertically.

Write the polynomials vertically.

$$2x^3 + 7x^2 + 5x - 1$$
$$\underline{\qquad\qquad 4x + 6}$$

Begin by multiplying each term in the top row by 6.

$$2x^3 + 7x^2 + 5x - 1$$
$$\underline{\qquad\qquad 4x + 6}$$
$$12x^3 + 42x^2 + 30x - 6$$

Now multiply each term in the top row by $4x$. Then add like terms.

$$2x^3 + 7x^2 + 5x - 1$$
$$\underline{\qquad\qquad 4x + 6}$$
$$12x^3 + 42x^2 + 30x - 6$$
$$\underline{8x^4 + 28x^3 + 20x^2 - 4x\qquad}$$
$$8x^4 + 40x^3 + 62x^2 + 26x - 6$$

The product is $8x^4 + 40x^3 + 62x^2 + 26x - 6$.

4. Multiply
$(4x^3 - 3x^2 + 6x + 5)(7x - 3)$
vertically.

_____

Objective 3 Practice Exercises

For extra help, see Examples 3–4 on page 390 of your text.

Find each product.

7. $(x+3)(x^2-3x+9)$

7. _____

8. $(2m^2+1)(3m^3+2m^2-4m)$

8. _____

9. $(3x^2+x)(2x^2+3x-4)$

9. _____

Objective 4   Multiply binomials using the FOIL method.

Video Examples

Review these examples for Objective 4:

5.   Use the FOIL method to find the product $(x+7)(x-5)$.

   Step 1 F Multiply the first terms: $x(x)=x^2$.

   Step 2 O Find the outer product: $x(-5)=-5x$.

   Step 3 I   Find the inner product: $7(x)=7x$.
   Add the outer and inner products mentally:
   $$-5x+7x=2x$$

   Step 4 L Multiply the last terms: $7(-5)=-35$.

   The product $(x+7)(x-5)$ is $x^2+2x-35$.

Now Try:

5.   Use the FOIL method to find the product $(x+9)(x-6)$.

   _____

6. Multiply $(7x-3)(4y+5)$.

First $7x(4y)=28xy$

Outer $7x(5)=35x$

Inner $-3(4y)=-12y$

Last $-3(5)=-15$

The product $(7x-3)(4y+5)$ is
$28xy+35x-12y-15$.

6. Multiply $(8y-7)(2x+9)$.

_____

7b. Find each product.
$(8p-5q)(8p+q)$

$(8p-5q)(8p+q)$
$=64p^2+8pq-40pq-5q^2$
$=64p^2-32pq-5q^2$

7b. Find each product.
$(9p+4q)(5p-q)$

_____

Objective 3 Practice Exercises

For extra help, see Examples 5–7 on page 392 of your text.

Find each product.

10. $(5a-b)(4a+3b)$

10. _____

11. $(3+4a)(1+2a)$

11. _____

12. $(2m+3n)(-3m+4n)$

12. _____

Chapter 5 EXPONENTS AND POLYNOMIALS

5.6   Special Products

Learning Objectives
1   Square binomials.
2   Find the product of the sum and difference of two terms.
3   Find greater powers of binomials.

Key Terms

Use the vocabulary terms listed below to complete each statement in exercises 1–2.

conjugate   **binomial**

1.   A polynomial with two terms is called a _____.

2.   The _____ of $a + b$ is $a - b$.

Objective 1   Square binomials.

Video Examples

Review these examples for Objective 1:

1.   Find $(m+5)^2$.

$$(m+5)^2 = (m+5)(m+5)$$
$$= m^2 + 5m + 5m + 25$$
$$= m^2 + 10m + 25$$

2.   Square each binomial.

b. $(7z-4)^2$

$$(7z-4)^2 = (7z)^2 - 2(7z)(4) + (-4)^2$$
$$= 7^2 z^2 - 56z + 16$$
$$= 49z^2 - 56z + 16$$

c. $(6x+3y)^2$

$$(6x+3y)^2 = (6x)^2 + 2(6x)(3y) + (3y)^2$$
$$= 36x^2 + 36xy + 9y^2$$

Now Try:

1.   Find $(x+6)^2$.

_____

2.   Square each binomial.

b. $(8a-3b)^2$

_____

c. $(2a+9k)^2$

_____

Objective 1 Practice Exercises

For extra help, see Examples 1–2 on pages 396–397 of your text.

Find each square by using the pattern for the square of a binomial.

1. $(7+x)^2$

1. _____

2. $(2m-3p)^2$

2. _____

3. $(4y-0.7)^2$

3. _____

Objective 2   **Find the product of the sum and difference of two terms.**

Video Examples

Review these examples for Objective 2:

3a. Find the product.

$$(x+5)(x-5)$$

Use the rule for the product of the sum and difference of two terms.

$$(x+5)(x-5) = x^2 - 5^2$$
$$= x^2 - 25$$

4. Find each product.

c. $\left(z-\dfrac{6}{7}\right)\left(z+\dfrac{6}{7}\right)$

$$\left(z-\dfrac{6}{7}\right)\left(z+\dfrac{6}{7}\right) = z^2 - \left(\dfrac{6}{7}\right)^2$$
$$= z^2 - \dfrac{36}{49}$$

b. $(6x+w)(6x-w)$

$$(6x+w)(6x-w) = (6x)^2 - w^2$$
$$= 36x^2 - w^2$$

Now Try:

3a. Find the product.

$$(x+9)(x-9)$$

_____

4. Find each product.

c. $\left(x+\dfrac{3}{5}\right)\left(x-\dfrac{3}{5}\right)$

_____

b. $(11x-y)(11x+y)$

_____

d. $3q(q^2+4)(q^2-4)$

First, multiply the conjugates.

$3q(q^2+4)(q^2-4)=3q(q^4-16)$

$\qquad\qquad\qquad = 3q^5-48q$

d. $4p(p^2+6)(p^2-6)$

_____

Objective 2 Practice Exercises

For extra help, see Examples 3–4 on pages 398–399 of your text.

Find each product by using the pattern for the sum and difference of two terms.

4. $(12+x)(12-x)$

4. _____

5. $(8k+5p)(8k-5p)$

5. _____

6. $\left(\frac{4}{7}t+2u\right)\left(\frac{4}{7}t-2u\right)$

6. _____

Objective 3   Find greater powers of binomials.

Video Examples

Review these examples for Objective 3:

5.   Find each product.

   a. $(x+4)^3$

   $(x+4)^3$

   $=(x+4)^2(x+4)$

   $=(x^2+8x+16)(x+4)$

   $=x^3+8x^2+16x+4x^2+32x+64$

   $=x^3+12x^2+48x+64$

Now Try:

5.   Find each product.

   a. $(x+6)^3$

   _____

Copyright © 2018 Pearson Education, Inc.

b. $(5y-4)^4$ **b.** $(3x-5)^4$

$(5y-4)^4$

$=(5y-4)^2(5y-4)^2$ _____

$=(25y^2-40y+16)(25y^2-40y+16)$

$=625y^4-1000y^3+400y^2$

$\quad -1000y^3+1600y^2-640y$

$\quad +400y^2-640y+256$

$=625y^4-2000y^3+2400y^2-1280y+256$

Objective 3 Practice Exercises

For extra help, see Example 5 on page 399 of your text.

Find each product.

7. $(a-3)^3$ 7. _____

8. $(j+3)^4$ 8. _____

9. $(4s+3t)^4$ 9. _____

Chapter 5 EXPONENTS AND POLYNOMIALS

5.7   Dividing a Polynomial by a Monomial

Learning Objectives
1   Divide a polynomial by a monomial.

Key Terms

Use the vocabulary terms listed below to complete each statement in exercises 1–3.

quotient   dividend   divisor

1.   In the division $\dfrac{5x^5 - 10x^3}{5x^2} = x^3 - 2x$, the expression $5x^5 - 10x^3$ is the

_____.

2.   In the division $\dfrac{5x^5 - 10x^3}{5x^2} = x^3 - 2x$, the expression $x^3 - 2x$ is the _____.

3.   In the division $\dfrac{5x^5 - 10x^3}{5x^2} = x^3 - 2x$, the expression $5x^2$ is the _____.

Objective 1   Divide a polynomial by a monomial.

Video Examples

Review these examples for Objective 1:

1.   Divide $6x^4 - 18x^3$ by $6x^2$.

$$\frac{6x^4 - 18x^3}{6x^2} = \frac{6x^4}{6x^2} - \frac{18x^3}{6x^2}$$
$$= x^2 - 3x$$

Check Multiply. $6x^2(x^2 - 3x) = 6x^4 - 18x^3$

2.   Divide. $\dfrac{25a^6 - 15a^4 + 10a^2}{5a^3}$

Divide each term by $5a^3$.

$$\frac{25a^6 - 15a^4 + 10a^2}{5a^3} = \frac{25a^6}{5a^3} - \frac{15a^4}{5a^3} + \frac{10a^2}{5a^3}$$
$$= 5a^3 - 3a + \frac{2}{a}$$

Now Try:

1.   Divide $6x^4 - 18x^3$ by $6x^2$.

_____

2.   Divide. $\dfrac{27n^5 - 36n^4 - 18n^2}{9n^3}$

_____

Check $5a^3\left(5a^3 - 3a + \dfrac{2}{a}\right)$

$\qquad = 5a^3\left(5a^3\right) + 5a^3\left(-3a\right) + 5a^3\left(\dfrac{2}{a}\right)$

$\qquad = 25a^6 - 15a^4 + 10a^2$

3. Divide $-12x^4 + 15x^5 - 5x$ by $-5x$.

Write the polynomial in descending powers before dividing.

$$\dfrac{15x^5 - 12x^4 - 5x}{-5x} = \dfrac{15x^5}{-5x} - \dfrac{12x^4}{-5x} - \dfrac{5x}{-5x}$$

$$= -3x^4 + \dfrac{12}{5}x^3 + 1$$

Check $-5x\left(-3x^4 + \dfrac{12}{5}x^3 + 1\right)$

$\qquad = -5x\left(-3x^4\right) - 5x\left(\dfrac{12}{5}x^3\right) - 5x(1)$

$\qquad = 15x^5 - 12x^4 - 5x$

4. Divide
$225x^5y^9 - 150x^3y^7 + 110x^2y^5 - 80xy^3 + 75y^2$
by $-25xy^2$.

$$\dfrac{225x^5y^9 - 150x^3y^7 + 110x^2y^5 - 80xy^3 + 75y^2}{-25xy^2}$$

$$= \dfrac{225x^5y^9}{-25xy^2} - \dfrac{150x^3y^7}{-25xy^2} + \dfrac{110x^2y^5}{-25xy^2} - \dfrac{80xy^3}{-25xy^2} + \dfrac{75y^2}{-25xy^2}$$

$$= -9x^4y^7 + 6x^2y^5 - \dfrac{22xy^3}{5} + \dfrac{16y}{5} - \dfrac{3}{x}$$

Check by multiplying the quotient by the divisor.

3. Divide $-8z^5 + 7z^6 - 10z - 6$ by $2z^2$.

_____

4. Divide
$80a^5b^3 + 160a^4b^2 - 120a^2b$ by
$-40a^2b$.

_____

223

Name: Date:

Instructor: Section:

Objective 1 Practice Exercises

For extra help, see Examples 1–4 on pages 403–404 of your text.

Perform each division.

1. $\dfrac{16a^5 - 24a^3}{8a^2}$

1. _____

2. $\dfrac{12z^5 + 28z^4 - 8z^3 + 3z}{4z^3}$

2. _____

3. $\dfrac{39m^4 - 12m^3 + 15}{-3m^2}$

3. _____

Chapter 5 EXPONENTS AND POLYNOMIALS

5.8   Dividing a Polynomial by a Polynomial

Learning Objectives
1   Divide a polynomial by a polynomial.
2   Apply polynomial division to a geometry problem.

Key Terms

Use the vocabulary terms listed below to complete each statement in exercises 1−3.

> **quotient**   **dividend**   **divisor**

1.   In the division $\dfrac{6x^2 - 9x - 12}{2x - 5} = 3x + 3 + \dfrac{3}{2x - 5}$, the expression $2x - 5$ is the

     _____.

2.   In the division $\dfrac{6x^2 - 9x - 12}{2x - 5} = 3x + 3 + \dfrac{3}{2x - 5}$, the expression $3x + 3 + \dfrac{3}{2x - 5}$ is

     the _____.

3.   In the division $\dfrac{6x^2 - 9x - 12}{2x - 5} = 3x + 3 + \dfrac{3}{2x - 5}$, the expression $6x^2 - 9x - 12$ is the

     _____.

Objective 1   Divide a polynomial by a polynomial.

Video Examples

Review these examples for Objective 1:

2.   Divide $8x + 9x^3 - 7 - 9x^2$ by $3x - 1$.

Write the dividend in descending powers as
$9x^3 - 9x^2 + 8x - 7$.

Step 1 $9x^3$ divided by $3x$ is $3x^2$.
$3x^2(3x - 1) = 9x^3 - 3x^2$

Step 2   Subtract. Bring down the next term.

Step 3 $-6x^2$ divided by $3x$ is $-2x$.
$-2x(3x - 1) = -6x^2 + 2x$

Step 4   Subtract. Bring down the next term.

Step 5 $6x$ divided by $3x$ is 2.
$2(3x - 1) = 6x - 2$

Now Try:

2.   Divide $-12x^2 + 10x^3 - 3 - 8x$
     by $5x - 1$.

     _____

$$3x - 1 \overline{) \begin{array}{r} 3x^2 - 2x + 2 \\ 9x^3 - 9x^2 + 8x - 7 \\ \underline{9x^3 - 3x^2} \\ -6x^2 + 8x \\ \underline{-6x^2 + 2x} \\ 6x - 7 \\ \underline{6x - 2} \\ -5 \end{array}}$$

$$\frac{9x^3 - 9x^2 + 8x - 7}{3x - 1} = 3x^2 - 2x + 2 + \frac{-5}{3x - 1}$$

Step 7 Multiply to check.

Check $(3x - 1)\left(3x^2 - 2x + 2 + \dfrac{-5}{3x - 1}\right)$

$$= (3x - 1)(3x^2) + (3x - 1)(-2x)$$

$$+ (3x - 1)(2) + (3x - 1)\left(\frac{-5}{3x - 1}\right)$$

$$= 9x^3 - 3x^2 - 6x^2 + 2x + 6x - 2 - 5$$

$$= 9x^3 - 9x^2 + 8x - 7$$

3. Divide $x^3 - 64$ by $x - 4$.

Here the dividend is missing the x^2-term and the x-term. We use 0 as the coefficient for each missing term.

$$x - 4 \overline{) \begin{array}{r} x^2 + 4x + 16 \\ x^3 + 0x^2 + 0x - 64 \\ \underline{x^3 - 4x^2} \\ 4x^2 + 0x \\ \underline{4x^2 - 16x} \\ 16x - 64 \\ \underline{16x - 64} \\ 0 \end{array}}$$

The remainder is 0. The quotient is $x^2 + 4x + 16$.

Check $(x - 4)(x^2 + 4x + 16)$

$$= x^3 + 4x^2 + 16x - 4x^2 - 16x - 64$$

$$= x^3 - 64$$

3. Divide $x^3 - 1000$ by $x - 10$.

_____

4. Divide $x^4 - 3x^3 + 7x^2 - 8x + 14$ by $x^2 + 2$.

Since $x^2 + 2$ is missing the x-term, we write it as $x^2 + 0x + 2$.

$$
\begin{array}{r}
x^2 - 3x + 5 \\
x^2 + 0x + 2 \overline{)\, x^4 - 3x^3 + 7x^2 - 8x + 14} \\
\underline{x^4 + 0x^3 + 2x^2} \\
-3x^3 + 5x^2 - 8x \\
\underline{-3x^3 + 0x^2 - 6x} \\
5x^2 - 2x + 14 \\
\underline{5x^2 + 0x + 10} \\
-2x + 4
\end{array}
$$

The quotient is $x^2 - 3x + 5 + \dfrac{-2x + 4}{x^2 + 2}$

The check shows that the quotient multiplied by the divisor gives the original dividend.

5. Divide $5x^3 + 8x^2 + 12x - 1$ by $5x + 5$.

$$
\begin{array}{r}
x^2 + \frac{3}{5}x + \frac{9}{5} \\
5x + 5 \overline{)\, 5x^3 + 8x^2 + 12x - 1} \\
\underline{5x^3 + 5x^2} \\
3x^2 + 12x \\
\underline{3x^2 + 3x} \\
9x - 1 \\
\underline{9x + 9} \\
-10
\end{array}
$$

The answer is $x^2 + \dfrac{3}{5}x + \dfrac{9}{5} - \dfrac{10}{5x + 5}$.

4. Divide
$3x^4 + 5x^3 - 7x^2 - 12x + 9$ by $x^2 - 4$.

_____

5. Divide $8x^3 - 7x^2 + 4x + 1$ by $8x - 8$.

_____

Objective 1 Practice Exercises

For extra help, see Examples 1–5 on pages 408–411 of your text.

Perform each division.

1. $\dfrac{-6x^2 + 23x - 20}{2x - 5}$

1. _____

2. $\dfrac{6x^4 - 12x^3 + 13x^2 - 5x - 1}{2x^2 + 3}$

2. _____

3. $\dfrac{2a^4 + 5a^2 + 3}{2a^2 + 3}$

3. _____

Copyright © 2018 Pearson Education, Inc.

Name: _____   Date: _____

Instructor: _____   Section: _____

Objective 2   Apply polynomial division to a geometry problem.

Video Examples

Review this example for Objective 2:

6. The area of a rectangle is given by $12p^3 - 7p^2 + 5p - 1$ square units, and the width is $4p - 1$ units. What is the length of the rectangle?

For a rectangle, $A = LW$. Solving for L gives $L = \frac{A}{W}$. Divide the area, $12p^3 - 7p^2 + 5p - 1$ by the width $4p - 1$.

$$
\begin{array}{r}
3p^2 - p + 1 \\
4p-1 \overline{\smash{\big)}\ 12p^3 - 7p^2 + 5p - 1} \\
\underline{12p^3 - 3p^2} \\
-4p^2 + 5p \\
\underline{-4p^2 + p} \\
4p - 1 \\
\underline{4p - 1} \\
0
\end{array}
$$

The length is $3p^2 - p + 1$ units.

Now Try:

6. The area of a rectangle is given by $6r^3 - 5r^2 + 16r - 5$ square units, and the width is $3r - 1$ units. What is the length of the rectangle?

_____

Objective 2 Practice Exercises

For extra help, see Example 6 on page 411 of your text.

Work each problem.

4. The area of a parallelogram is given by $4y^3 - 44y - 600$ square units, and the height is $y - 6$ units. What is the base of the parallelogram?

4. _____

5. The area of a parallelogram is given by   5. _____
    $3t^3 + 16t^2 - 32t - 64$ square units, and the base is
    $t^2 + 4t - 16$ units. What is the height of the
    parallelogram?

6. The area of a rectangle is given by   6. _____
    $20x^3 - 19x^2 + 8x - 1$ square units, and the width is
    $5x - 1$ units. What is the length of the rectangle?

Chapter 6 FACTORING AND APPLICATIONS

6.1   Greatest Common Factors; Factor by Grouping

Learning Objectives
1   Find the greatest common factor of a list of numbers.
2   Find the greatest common factor of a list of variable terms.
3   Factor out the greatest common factor.
4   Factor by grouping.

Key Terms

Use the vocabulary terms listed below to complete each statement in exercises 1−4.

> **factor**   **factored form**   **greatest common factor (GCF)**
>
> **factoring**

1.   The process of writing a polynomial as a product is called _____.

2.   An expression is in _____ when it is written as a product.

3.   The _____ is the largest quantity that is a factor of each
      of a group of quantities.

4.   An expression A is a _____ of an expression B if B can be divided
      by A with 0 remainder.

Objective 1   Find the greatest common factor of a list of numbers.

Video Examples

Review these examples for Objective 1:

1.   Find the greatest common factor for each list of
      numbers.

      b. 90, 36, 108

      Write the prime factored form of each number.
      $$90 = 2 \cdot 3 \cdot 3 \cdot 5$$
      $$36 = 2 \cdot 2 \cdot 3 \cdot 3$$
      $$108 = 2 \cdot 2 \cdot 3 \cdot 3 \cdot 3$$
      There is one factor of 2 and two factors of 3.
      $$GCF = 2^1 \cdot 3^2 = 18$$

Now Try:

1.   Find the greatest common factor
      for each list of numbers.

      b. 32, 40, 72

      _____

c. 17, 18, 24

Write the prime factored form of each number.

$17 = 17$

$18 = 2 \cdot 3 \cdot 3$

$24 = 2 \cdot 2 \cdot 2 \cdot 3$

There are no primes common to all three numbers, so the GCF is 1.

c. 26, 27, 28

_____

Objective 1 Practice Exercises

For extra help, see Example 1 on pages 426–427 of your text.

Find the greatest common factor for each group of numbers.

1.   84, 280, 112

1. _____

2.   56, 21, 49

2. _____

3.   42, 48, 72

3. _____

Objective 2   **Find the greatest common factor of a list of variable terms.**

Video Examples

Review these examples for Objective 2:	Now Try:
2. Find the greatest common factor for each list of terms.	**2.** Find the greatest common factor for each list of terms.

a. $28m^4$, $35m^6$, $49m^9$, $70m^5$ **a.** $54x^5$, $48x^7$, $42x^9$, $30x^4$

$$28m^4 = 2 \cdot 2 \cdot 7 \cdot m^4$$
$$35m^6 = 5 \cdot 7 \cdot m^6$$
$$49m^9 = 7 \cdot 7 \cdot m^9$$
$$70m^5 = 2 \cdot 5 \cdot 7 \cdot m^5$$

Then, GCF $= 7m^4$.

_____

b. a^5b^3, a^7b^4, a^3b^5, b^9 **b.** p^5q^7, p^4q^3, p^8q^5, q^9

$$a^5b^3 = a^5 \cdot b^3$$
$$a^7b^4 = a^7 \cdot b^4$$
$$a^3b^5 = a^3 \cdot b^5$$
$$b^9 = b^9$$

Then, GCF $= b^3$.

_____

Objective 2 Practice Exercises

For extra help, see Example 2 on page 428 of your text.

Find the greatest common factor for each list of terms.

     4. $12ab^3$, $18a^2b^4$, $26ab^2$, $32a^2b^2$ **4.** _____

     5. $6k^2m^4n^5$, $8k^3m^7n^4$, $k^4m^8n^7$ **5.** _____

6. $9xy^4,\ 72x^4y^7,\ 27xy^2,\ 108x^2y^5$ **6.** _____

Objective 3   Factor out the greatest common factor.

Video Examples

Review these examples for Objective 3:	**Now Try:**
3. Write in factored form by factoring out the greatest common factor.	**3.** Write in factored form by factoring out the greatest common factor.

a. $7a^2 + 14a$

$\text{GCF} = 7a$

$$7a^2 + 14a = 7a(a) + 7a(2)$$
$$= 7a(a+2)$$

Check Multiply the factored form.
$$7a(a+2) = 7a(a) + 7a(2)$$
$$= 7a^2 + 14a$$

a. $8x^5 + 24x$

_____

b. $12x^5 + 27x^4 - 15x^3$

$\text{GCF} = 3x^3$

$$12x^5 + 27x^4 - 15x^3$$
$$= 3x^3(4x^2) + 3x^3(9x) + 3x^3(-5)$$
$$= 3x^3(4x^2 + 9x - 5)$$

Check Multiply the factored form.
$$3x^3(4x^2 + 9x - 5)$$
$$= 3x^3(4x^2) + 3x^3(9x) + 3x^3(-5)$$
$$= 12x^5 + 27x^4 - 15x^3$$

b. $20y^4 - 12y^3 + 4y^2$

_____

5a. Write in factored form by factoring out the greatest common factor.

$$x(x+9) + 7(x+9)$$

Factor out $x+9$.
$$x(x+9) + 7(x+9) = (x+9)(x+7)$$

5a. Write in factored form by factoring out the greatest common factor.

$$y(y+8) + 4(y+8)$$

_____

Objective 3 Practice Exercises

For extra help, see Examples 3–5 on pages 428–430 of your text.

Factor out the greatest common factor or a negative common factor if the coefficient of the term of greatest degree is negative.

7. $20x^2 + 40x^2 y - 70xy^2$ 7. _____

8. $2a(x - 2y) + 9b(x - 2y)$ 8. _____

9. $26x^8 - 13x^{12} + 52x^{10}$ 9. _____

Objective 4   Factor by grouping.

Video Examples

Review these examples for Objective 4:

6. Factor by grouping.

   a. $5x + 15 + bx + 3b$

   Group the first two terms and the last two terms.
   $5x + 15 + bx + 3b$
   $$= (5x + 15) + (bx + 3b)$$
   $$= 5(x + 3) + b(x + 3)$$
   $$= (x + 3)(5 + b)$$

   Check   Use the FOIL method.
   $$(x + 3)(5 + b) = 5x + bx + 3(5) + 3b$$
   $$= 5x + 15 + bx + 3b$$

Now Try:

6. Factor by grouping.

   a. $2x + xy + 14 + 7y$

   _____

c. $3x^2 - 18x + 5xy - 30y$

$3x^2 - 18x + 5xy - 30y$

$= (3x^2 - 18x) + (5xy - 30y)$

$= 3x(x-6) + 5y(x-6)$

$= (x-6)(3x+5y)$

Check by multiplying using the FOIL method.

c. $4x^2 - 28x + 5xy - 35y$

_____

Objective 4 Practice Exercises

For extra help, see Examples 6–7 on pages 430–432 of your text.

Factor each polynomial by grouping.

10. $15 - 5x - 3y + xy$

10. _____

11. $2x^2 - 14xy + xy - 7y^2$

11. _____

12. $3r^3 - 2r^2s + 3s^2r - 2s^3$

12. _____

Chapter 6 FACTORING AND APPLICATIONS

6.2   Factoring Trinomials

Learning Objectives
1   Factor trinomials with a coefficient of 1 for the second-degree term.
2   Factor such trinomials after factoring out the greatest common factor.

Key Terms

Use the vocabulary terms listed below to complete each statement in exercises 1−3.

prime polynomial   factoring   greatest common factor

1.   _____ is the process of writing a polynomial as a product.

2.   The _____ of a polynomial is the greatest term that is a factor of all the terms in the polynomial.

3.   A _____ is a polynomial that cannot be factored using only integers.

Objective 1   Factor trinomials with a coefficient of 1 for the second-degree term.

Video Examples

Review these examples for Objective 1:

2.   Factor $x^2 - 11x + 28$.

Look for integers whose product is 28 and whose sum is −11. Since the numbers have a positive product and a negative sum, we consider only pairs of negative integers.

Factors of 28	Sums of Factors
$-28, -1$	$-28 + (-1) = -29$
$-14, -2$	$-14 + (-2) = -16$
$-7, -4$	$-7 + (-4) = -11$

The required integers are −7 and −4.

$x^2 - 11x + 28$ factors as $(x - 7)(x - 4)$

Check   Use the FOIL method.

$$(x - 7)(x - 4) = x^2 - 4x - 7x + 28$$
$$= x^2 - 11x + 28$$

Now Try:

2.   Factor $y^2 - 12y + 35$.

_____

Name: Date:

Instructor: Section:

4.   Factor $x^2 - 10x - 39$.

     Look for integers whose product is –39 and whose sum is –10. Because the constant term, –39, is negative, we need pairs of integers with different signs.

Factors of -39	Sums of Factors
$39, -1$	$39 + (-1) = 38$
$-39, 1$	$-39 + 1 = -38$
$3, -13$	$3 + (-13) = -10$

The required integers are –13 and 3.

     $x^2 - 10x - 39$ factors as $(x - 13)(x + 3)$

Check   Use the FOIL method.

$$(x - 13)(x + 3) = x^2 + 3x - 13x - 39$$
$$= x^2 - 10x - 39$$

5a. Factor the trinomial.

     $x^2 - 7x + 18$

     Look for integers whose product is 18 and whose sum is –7. Since the numbers have a positive product and a negative sum, we consider only pairs of negative integers.

Factors of 18	Sums of Factors
$-18, -1$	$-18 + (-1) = -19$
$-9, -2$	$-9 + (-2) = -11$
$-6, -3$	$-6 + (-3) = -9$

None of the pairs of integers has a sum of –7.

     $x^2 - 7x + 18$ cannot be factored.

It is a prime polynomial.

6.   Factor $x^2 - 6xy - 7y^2$.

     Here, the coefficient of x in the middle term is –6y, so we need to find two expressions whose product is $-7y^2$ and whose sum is –6y.

Factors of $-7y^2$	Sums of Factors
$7y, -y$	$7y + (-y) = 6y$
$-7y, y$	$-7y + y = -6y$

     $x^2 - 6xy - 7y^2$ factors as $(x - 7y)(x + y)$

4.   Factor $a^2 - 15a - 34$.

              _____

5a. Factor the trinomial.

     $m^2 - 7m + 5$

              _____

6.   Factor $p^2 - 5pq - 14q^2$.

              _____

Check   Use the FOIL method.

$$(x-7y)(x+y) = x^2 + xy - 7xy + 7y^2$$
$$= x^2 - 6xy - 7y^2$$

Objective 1 Practice Exercises

For extra help, see Examples 1–6 on pages 437–439 of your text.

Factor completely. If a polynomial cannot be factored, write prime.

1. $r^2 + r + 3$ 1. _____

2. $x^2 - 11x + 28$ 2. _____

3. $x^2 - 8x - 33$ 3. _____

Objective 2   Factor such trinomials after factoring out the greatest common factor.

Video Examples

Review this example for Objective 2:	**Now Try:**

Review this example for Objective 2:

7.   Factor $5x^5 - 45x^4 + 90x^3$.

There is no second-degree term. Look for a common factor.

$$5x^5 - 45x^4 + 90x^3 = 5x^3\left(x^2 - 9x + 18\right)$$

Now factor $x^2 - 9x + 18$. The integers -3 and -6 have a product of 18 and a sum of -9.

$5x^5 - 45x^4 + 90x^3$ factors as $5x^3(x-6)(x-3)$

Check   Use the FOIL method.

$5x^3(x-6)(x-3)$

$$= 5x^3\left(x^2 - 3x - 6x + 18\right)$$

$$= 5x^3\left(x^2 - 9x + 18\right)$$

$$= 5x^5 - 45x^4 + 90x^3$$

Now Try:

7.   Factor $7x^6 - 49x^5 + 70x^4$.

_____

Objective 2 Practice Exercises

For extra help, see Example 7 on page 440 of your text.

Factor completely. If a polynomial cannot be factored, write **prime**.

4. $2n^4 - 16n^3 + 30n^2$

4. _____

5. $2a^3b - 10a^2b^2 + 12ab^3$

5. _____

6. $10k^6 + 70k^5 + 100k^4$

6. _____

Chapter 6 FACTORING AND APPLICATIONS

6.3   Factoring Trinomials by Grouping

Learning Objectives
1   Factor trinomials by grouping when the coefficient of the second-degree term is not 1.

Key Terms

Use the vocabulary terms listed below to complete each statement in exercises 1−2.

   coefficient   trinomial

1.   In the term $6x^2y$, 6 is the _____.

2.   A polynomial with three terms is a _____.

Objective 1   Factor trinomials by grouping when the coefficient of the second-degree term is not 1.

Video Examples

Review these examples for Objective 1:

2.   Factor each trinomial.

   a. $8x^2 - 2x - 1$

   We must find two integers with a product of $8(-1) = -8$ and a sum of –2. The integers are –4 and 2. We write the middle term as $-4x + 2x$.

   $$8x^2 - 2x - 1 = 8x^2 - 4x + 2x - 1$$
   $$= (8x^2 - 4x) + (2x - 1)$$
   $$= 4x(2x - 1) + 1(2x - 1)$$
   $$= (2x - 1)(4x + 1)$$

   Check   Multiply $(2x - 1)(4x + 1)$ to obtain $8x^2 - 2x - 1$.

   b. $15z^2 + z - 2$

   Look for two integers whose product is $15(-2) = -30$ and whose sum is 1.

   The integers are 6 and –5.

Now Try:

2.   Factor each trinomial.

   a. $14x^2 - 3x - 5$

   _____

   b. $3m^2 - m - 14$

   _____

$$15z^2 + z - 2 = 15z^2 - 5z + 6z - 2$$
$$= (15z^2 - 5z) + (6z - 2)$$
$$= 5z(3z - 1) + 2(3z - 1)$$
$$= (3z - 1)(5z + 2)$$

Check Multiply $(3z - 1)(5z + 2)$ to obtain
$15z^2 + z - 2$.

c. $12r^2 + 5rs - 2s^2$

Two integers whose product is $12(-2) = -24$
and whose sum is 5 are 8 and -3. Rewrite the
trinomial with four terms.
$$12r^2 + 5rs - 2s^2 = 12r^2 + 8rs - 3rs - 2s^2$$
$$= (12r^2 + 8rs) + (-3rs - 2s^2)$$
$$= 4r(3r + 2s) - s(3r + 2s)$$
$$= (3r + 2s)(4r - s)$$

Check Multiply $(3r + 2s)(4r - s)$ to obtain
$12r^2 + 5rs - 2s^2$.

3. Factor $100x^5 + 140x^4 - 15x^3$.

Factor out the greatest common factor, $5x^3$.
$$100x^5 + 140x^4 - 15x^3 = 5x^3(20x^2 + 28x - 3)$$

To factor $20x^2 + 28x - 3$, find two integers
whose product is $20(-3) = -60$ and whose sum
is 28. Factor 60 into prime factors.
$$60 = 2 \cdot 2 \cdot 3 \cdot 5$$
Combine the prime factors in pairs using
one positive factor and one negative factor
to get -60. The factors of 30 and -2 have the
correct sum, 28.
$$100x^5 + 140x^4 - 15x^3$$
$$= 5x^3(20x^2 + 28x - 3)$$
$$= 5x^3(20x^2 + 30x - 2x - 3)$$
$$= 5x^3[(20x^2 + 30x) + (-2x - 3)]$$
$$= 5x^3[10x(2x + 3) - 1(2x + 3)]$$
$$= 5x^3(2x + 3)(10x - 1)$$

c. $10x^2 + xy - 3y^2$

_____

3. Factor $30x^5 + 87x^4 - 63x^3$.

_____

Name: _____   Date: _____

Instructor: _____   Section: _____

Objective 1 Practice Exercises

For extra help, see Examples 1–3 on page 443–444 of your text.

Factor each trinomial by grouping.

1. $8b^2 + 18b + 9$

1. _____

2. $7a^2b + 18ab + 8b$

2. _____

3. $10c^2 - 29ct + 21t^2$

3. _____

Chapter 6 FACTORING AND APPLICATIONS

6.4   Factoring Trinomials Using the FOIL Method

Learning Objectives
1   Factor trinomials using the FOIL method.

Key Terms

Use the vocabulary terms listed below to complete each statement in exercises 1–3.

> **FOIL**   **outer product**   **inner product**

1. The _____ of $(2y-5)(y+8)$ is $-5y$.

2. _____ is a shortcut method for finding the product of two binomials.

3. The _____ of $(2y-5)(y+8)$ is $16y$.

Objective 1   Factor trinomials using the FOIL method.

Video Examples

Review these examples for Objective 1:

1. Factor $5x^2 + 14x + 8$.

    The product of the first terms of the binomial is $5x^2$. The possible factors are $5x$ and x, since we consider only positive factors. So we have

    $$5x^2 + 14x + 8 = (5x + \underline{\hphantom{00}})(x + \underline{\hphantom{00}})$$

    The product of the last terms is 8. The positive factors are $1 \cdot 8$, $8 \cdot 1$, $4 \cdot 2$, or $2 \cdot 4$. We want the middle term of 14.

    Try $1 \cdot 8$ in $(5x + \underline{\hphantom{00}})(x + \underline{\hphantom{00}})$.

    $$(5x + 1)(x + 8)$$

    gives middle term $40x + x = 41x$. Incorrect.

    Try $8 \cdot 1$ in $(5x + \underline{\hphantom{00}})(x + \underline{\hphantom{00}})$.

    $$(5x + 8)(x + 1)$$

    gives middle term $5x + 8x = 13x$. Incorrect.

    Try $4 \cdot 2$ in $(5x + \underline{\hphantom{00}})(x + \underline{\hphantom{00}})$.

    $$(5x + 4)(x + 2)$$

    gives middle term $10x + 4x = 14x$. Correct.

    Thus, $5x^2 + 14x + 8 = (5x + 4)(x + 2)$.

    Check. Multiply $(5x + 4)(x + 2)$ to obtain $5x^2 + 14x + 8$.

Now Try:

1. Factor $3x^2 + 17x + 10$.

    _____

2. Factor $6x^2 + 13x + 7$.

The number 6 has several possible pairs of factors, but 7 has only 1 and 7, or −1 and −7. We choose positive factors since all coefficients in the trinomial are positive.

$$(\underline{} + 7)(\underline{} + 1)$$

The possible pairs of $6x^2$ are $6x$ and x, or $3x$ and $2x$.

$$(3x + 7)(2x + 1)$$

gives middle term $3x + 14x = 17x$. Incorrect.

$$(2x + 7)(3x + 1)$$

gives middle term $2x + 21x = 23x$. Incorrect.

$$(6x + 7)(x + 1)$$

gives middle term $6x + 7x = 13x$. Correct.

$6x^2 + 13x + 7$ factors as $(6x + 7)(x + 1)$.

Check. Multiply $(6x + 7)(x + 1)$ to obtain $6x^2 + 13x + 7$.

3. Factor $10x^2 - 19x + 7$.

Since 7 has only 1 and 7 or −1 and −7 as factors, it is better to begin by factoring 7. We need two negative factors because the product of two negative factors is positive and their sum is negative, as required.
We try −1 and −7.

$$(\underline{} - 1)(\underline{} - 7)$$

The factors of $10x^2$ are $10x$ and x, or $5x$ and $2x$.

$$(10x - 1)(x - 7)$$

has middle term $-70x - x = -71x$. Incorrect.

$$(5x - 1)(2x - 7)$$

has middle term $-35x - 2x = -37x$. Incorrect.

$$(2x - 1)(5x - 7)$$

has middle term $-14x - 5x = -19x$. Correct.

Thus $10x^2 - 19x + 7$ factors as $(2x - 1)(5x - 7)$.

2. Factor $15x^2 + 26x + 7$.

_____

3. Factor $20x^2 - 13x + 2$.

_____

4. Factor $6x^2 - x - 15$.

The integer 6 has several possible pairs of factors, as does -15. Since the constant term is negative, one positive factor and one negative factor of -15 are needed. Since the coefficient of the middle term is relatively small, it is wise to avoid large factors. We try $3x$ and $2x$ as factors of $6x^2$ and 5 and -3 as factors of -15.

$(3x + 5)(2x - 3)$
has middle term $-9x + 10x = x$. Incorrect.
$(3x - 5)(2x + 3)$
has middle term $9x - 10x = -x$. Correct.

$6x^2 - x - 15$ factors as $(3x - 5)(2x + 3)$.

5. Factor $18x^2 - 3xy - 28y^2$.

There are several factors of $18x^2$, including
$\quad 18x$ and x, $9x$ and $2x$, and $6x$ and $3x$.
There are many possible pairs of factors of $-28y^2$, including
$\quad 28y$ and $-y$, $-28y$ and y, $14y$ and $-2y$,
$\quad -14y$ and $2y$, $7y$ and $-4y$, $-7y$ and $4y$.

Once again, since the coefficient of the middle term is relatively small, avoid the larger factors. Try the factors of $6x$ and $3x$, and $4y$ and $-7y$.
$(6x + 4y)(3x - 7y)$ Incorrect
The first binomial has a common factor of 2.
$(6x - 7y)(3x + 4y)$
has middle term $24xy - 21xy = 3xy$. Incorrect.
Interchange the signs of the last two terms.
$(6x + 7y)(3x - 4y)$
has middle term $-24xy + 21xy = -3xy$. Correct.

Thus, $18x^2 - 3xy - 28y^2$ factors as
$(6x + 7y)(3x - 4y)$

4. Factor $8x^2 + 2x - 21$.

_____

5. Factor $24x^2 - 2xy - 15y^2$.

_____

Name: Date:
Instructor: Section:

Objective 1 Practice Exercises

For extra help, see Examples 1–6 on pages 447–450 of your text.

Factor each trinomial completely.

1. $8q^2 + 10q + 3$ 1. _____

2. $3a^2 + 8ab + 4b^2$ 2. _____

3. $4c^2 + 14cd - 8d^2$ 3. _____

Chapter 6 FACTORING AND APPLICATIONS

6.5   Special Factoring Techniques

Learning Objectives
1   Factor a difference of squares.
2   Factor a perfect square trinomial.

Key Terms

Use the vocabulary terms listed below to complete each statement in exercises 1–2.

> **perfect square trinomial**   **difference**

1.   A _____ is the result of a subtraction.

2.   A _____ is a trinomial that can be factored as the square of a binomial.

Objective 1   Factor a difference of squares.

Video Examples

Review these examples for Objective 1:

1b. Factor each binomial, if possible.

$$y^2 - 27$$

Because 27 is not a square of an integer, this binomial is not a difference in squares. It is a prime polynomial.

2a. Factor each difference of squares.

$$36x^2 - 25$$

$$36x^2 - 25 = (6x)^2 - 5^2$$
$$= (6x + 5)(6x - 5)$$

3.   Factor completely.

   a. $28y^2 - 175$

$$28y^2 - 175 = 7(4y^2 - 25)$$
$$= 7\left[(2y)^2 - 5^2\right]$$
$$= 7(2y + 5)(2y - 5)$$

Now Try:

1b. Factor each binomial, if possible.
   $$y^2 - 110$$

   _____

2a. Factor each difference of squares.
   $$4x^2 - 81$$

   _____

3.   Factor completely.

   a. $90x^2 - 490$

   _____

b. $y^4 - 49$

$$y^4 - 49 = \left(y^2\right)^2 - 7^2$$
$$= \left(y^2 + 7\right)\left(y^2 - 7\right)$$

c. $p^4 - 81$

$$p^4 - 81 = \left(p^2\right)^2 - 9^2$$
$$= \left(p^2 + 9\right)\left(p^2 - 9\right)$$
$$= \left(p^2 + 9\right)\left(p + 3\right)\left(p - 3\right)$$

b. $x^4 - 64$

_____

c. $p^4 - 256$

_____

Objective 1 Practice Exercises

For extra help, see Examples 1–3 on pages 453–454 of your text.

Factor each binomial completely. If a binomial cannot be factored, write **prime**.

1. $x^2 - 49$ 1. _____

2. $81x^4 - 16$ 2. _____

3. $9x^2 + 16$ 3. _____

Objective 2   Factor a perfect square trinomial.

Video Examples

Review these examples for Objective 2:	**Now Try:**

Review these examples for Objective 2:

4.   Factor $x^2 + 20x + 100$.

The terms x^2 and 100 are perfect squares.
$$x^2 + 20x + 100 = (x+10)^2$$
Check the middle term. $2(x)(10) = 20x$
The trinomial is a perfect square.

5d. Factor each trinomial.

$$128z^3 + 192z^2 + 72z$$

$$128z^3 + 192z^2 + 72z$$
$$= 8z(16z^2 + 24z + 9)$$
$$= 8z\left[(4z)^2 + 2(4z)(3) + 3^2\right]$$
$$= 8z(4z+3)^2$$

Now Try:

4.   Factor $p^2 + 16p + 64$.

_____

5d. Factor each trinomial.

$$20x^3 + 100x^2 + 125x$$

_____

Objective 2 Practice Exercises

For extra help, see Examples 4–5 on pages 455–456 of your text.

Factor each trinomial completely. It may be necessary to factor out the greatest common factor first.

4. $z^2 - \dfrac{4}{3}z + \dfrac{4}{9}$

4. _____

5. $9j^2 + 12j + 4$

5. _____

6. $-12a^2 + 60ab - 75b^2$

6. _____

Chapter 6 FACTORING AND APPLICATIONS

6.6   Solving Quadratic Equations using the Zero-Factor Property

Learning Objectives
1   Solve quadratic equations using the zero-factor property.
2   Solve other equations using the zero-factor property.

Key Terms

Use the vocabulary terms listed below to complete each statement in exercises 1–2.

quadratic equation   standard form

1.   An equation written in the form $ax^2 + bx + c = 0$ is written in the
       _____ of a quadratic equation.

2.   An equation that can written in the form $ax^2 + bx + c = 0$, with $a \neq 0$, is a
       _____.

Objective 1   Solve quadratic equations using the zero-factor property.

Video Examples

Review these examples for Objective 1:
1a. Solve each equation.

$$(x+9)(5x-6)=0$$

By the zero-factor property, either $x+9=0$ or $5x-6=0$, or both.

$$x+9=0 \quad \text{or} \quad 5x-6=0$$
$$x=-9 \quad \text{or} \quad 5x=6$$
$$x=\frac{6}{5}$$

Check:
Let $x = -9$.
$$(x+9)(5x-6)=0$$
$$(-9+9)[5(-9)-6]\overset{?}{=}0$$
$$0(-51)\overset{?}{=}0$$
$$0=0 \;\; \text{True}$$

Now Try:
1a. Solve each equation.

$$(x+12)(4x-7)=0$$

_____

Let $x = \dfrac{6}{5}$.

$$(x+9)(5x-6)=0$$

$$\left(\dfrac{6}{5}+9\right)\left[5\left(\dfrac{6}{5}\right)-6\right] \overset{?}{=} 0$$

$$\left(\dfrac{51}{5}\right)(6-6) \overset{?}{=} 0$$

$$0 = 0 \quad \text{True}$$

Both values check, so the solution set is

$$\left\{-9,\ \dfrac{6}{5}\right\}.$$

2a. Solve each equation.

$$x^2 - 6x = -5$$

First, write the equation in standard form.

$$x^2 - 6x = -5$$

$$x^2 - 6x + 5 = 0$$

Factor and use the zero-factor property.

$$(x-1)(x-5)=0$$

$$x-1=0 \quad \text{or} \quad x-5=0$$

$$x=1 \quad \text{or} \qquad x=5$$

Check these solutions by substituting each in the original equation. The solution set is $\{1, 5\}$.

3. Solve $8p^2 + 30 = 46p$.

$$8p^2 + 30 = 46p$$

$$8p^2 - 46p + 30 = 0 \quad \text{Standard form}$$

$$2\left(4p^2 - 23p + 15\right)=0 \quad \text{Factor out 2.}$$

$$4p^2 - 23p + 15 = 0 \quad \text{Divide each side by 2}$$

$$(4p-3)(p-5)=0 \quad \text{Factor.}$$

$$4p-3=0 \quad \text{or} \quad p-5=0 \quad \text{Zero-factor}$$

$$p = \dfrac{3}{4} \quad \text{or} \qquad p=5 \quad \text{property}$$

The solution set is $\left\{\dfrac{3}{4},\ 5\right\}$.

2a. Solve each equation.

$$x^2 - 5x = 24$$

_____

3. Solve $15p^2 + 36 = 57p$.

_____

4. Solve each equation.

   c. $k(3k+5)=2$

$$k(3k+5)=2$$
$$3k^2+5k=2$$
$$3k^2+5k-2=0$$
$$(3k-1)(k+2)=0$$
$$3k-1=0 \quad \text{or} \quad k+2=0$$
$$3k=1 \quad \text{or} \quad k=-2$$
$$k=\frac{1}{3}$$

The solution set is $\left\{-2,\ \frac{1}{3}\right\}$.

   b. $y^2=9y$

$$y^2=9y$$
$$y^2-9y=0$$
$$y(y-9)=0$$
$$y=0 \quad \text{or} \quad y-9=0$$
$$y=9$$

The solution set is $\{0,\ 9\}$.

4. Solve each equation.

   c. $k(4k-23)=6$

_____

   b. $y^2=11y$

_____

Objective 1 Practice Exercises

For extra help, see Examples 1–5 on pages 464–467 of your text.

Solve each equation and check your solutions.

1. $2x^2-3x-20=0$ **1.** _____

2. $25x^2=20x$ **2.** _____

3. $c(5c+17)=12$ **3.** _____

Objective 2   Solve other equations using the zero-factor property.

Video Examples

Review this example for Objective 2:

6. Solve $12z^3 - 3z = 0$.

$$12z^3 - 3z = 0$$
$$3z(4z^2 - 1) = 0$$
$$3z(2z+1)(2z-1) = 0$$

By an extension of the zero-factor property, we have

$3z = 0$ or $2z+1=0$ or $2z-1=0$

$z = 0$ or $z = -\dfrac{1}{2}$ or $z = \dfrac{1}{2}$

Check by substituting each value in the original equation. The solution set is $\left\{-\dfrac{1}{2},\ 0,\ \dfrac{1}{2}\right\}$.

Now Try:

6. Solve $3r^3 = 75r$.

_____

Objective 2 Practice Exercises

For extra help, see Examples 6–7 on page 46 of your text.

Solve each equation and check your solutions.

4. $x^3 + 2x^2 - 8x = 0$ **4.** _____

5. $z^4 + 8z^3 - 9z^2 = 0$ **5.** _____

6. $\left(y^2 - 5y + 6\right)\left(y^2 - 36\right) = 0$ **6.** _____

Chapter 6 FACTORING AND APPLICATIONS

6.7   Applications of Quadratic Equations

Learning Objectives
1   Solve problems involving geometric figures.
2   Solve problems involving consecutive integers.
3   Solve problems by applying the Pythagorean theorem.
4   Solve problems using given quadratic models.

Key Terms

Use the vocabulary terms listed below to complete each statement in exercises 1–2.

hypotenuse   **legs**

1.   In a right triangle, the sides that form the right angle are the _____.

2.   The longest side of a right triangle is the _____.

Objective 1   Solve problems involving geometric figures.

Video Examples

Review this example for Objective 1:

1.   The length of a rectangle is three times its width. If the width was increased by 4 and the length remained the same, the resulting rectangle would have an area of 231 square inches. Find the dimensions of the original rectangle.

Step 1 Read the problem carefully. Find the dimensions of the original rectangle.

Step 2 Assign a variable.
Let x = width.
    $3x$ = length
  $x + 4$ = new width
    $3x$ = length

Step 3 Write an equation. The area of the rectangle is given by $\text{Area} = \text{Length} \times \text{Width}$. Substitute 231 for area, $3x$ for length, and $x + 4$ for width.

$$231 = 3x(x+4)$$

Now Try:

1.   Mr. Fixxall is building a box which will have a volume of 60 cubic meters. The height of the box will be 4 meters, and the length will be 2 meters more than the width. Find the width and length of the box.

_____

Step 4 Solve.
$$231 = 3x(x+4)$$
$$231 = 3x^2 + 12x$$
$$3x^2 + 12x - 231 = 0$$
$$3(x^2 + 4x - 77) = 0$$
$$x^2 + 4x - 77 = 0$$
$$(x+11)(x-7) = 0$$
$$x + 11 = 0 \quad \text{or} \quad x - 7 = 0$$
$$x = -11 \quad \text{or} \quad x = 7$$

Step 5 State the answer. The solutions are −11 and 7. A rectangle cannot have a side of negative length, so we discard −11. The width is 7 inches. The length is $3(7) = 21$ inches.

Step 6 Check. The new width is 7 + 4 = 11. The new area of the rectangle is $11(21) = 231$ square inches.

Objective 1 Practice Exercises

For extra help, see Example 1 on pages 472–473 of your text.

Solve each problem. Check your answers to be sure they are reasonable.

1. A book is three times as long as it is wide. Find the length and width of the book in inches if its area is numerically 128 more than its perimeter.

    1. width_____

       length _____

Name: Date:
Instructor: Section:

2. The area of a triangle is 42 square centimeters. The base is 2 centimeters less than twice the height. Find the base and height of the triangle.

2. base____________

height ___________

3. The volume of a box is 192 cubic feet. If the length of the box is 8 feet and the width is 2 feet more than the height, find the height and width of the box.

3. height ___________

width___________

Objective 2   Solve problems involving consecutive integers.

Video Examples

Review these examples for Objective 2:	**Now Try:**

Review these examples for Objective 2:

3. Find two consecutive positive even integers whose product is 168.

Step 1 Read carefully. Note that the integers are positive consecutive even integers.

Step 2 Assign a variable.
     Let x = the first integer.
     Then $x + 2$ = the next even integer.

Step 3 Write an equation. The product is 168.
$$x(x+2)=168$$

Step 4 Solve.
$$x^2 + 2x = 168$$
$$x^2 + 2x - 168 = 0$$
$$(x+14)(x-12)=0$$
$$x+14=0 \quad \text{or} \quad x-12=0$$
$$x=-14 \quad \text{or} \quad x=12$$

Step 5 State the answer. The solutions are –14 and 12. We discard –14 since the integers are required to be positive. If x = 12, then $x + 2$ = 14. So, the numbers are 12 and 14.

Step 6 Check. The numbers 12 and 14 are consecutive positive even integers, and their product is 168.

2. Find two consecutive integers such that the sum of the squares of the two integers is 3 more than the opposite (additive inverse) of the smaller integer.

Step 1 Read the problem. Note that the numbers are consecutive.

Step 2 Assign a variable.
     Let x = the first number.
     Then $x + 1$ = the second number.

Step 3 Write an equation. The sum of squares is three more than the opposite of the smaller integer.
$$x^2 + (x+1)^2 = 3 + (-x)$$

Now Try:

3. The product of two consecutive even positive integers is 10 more than seven times the larger. Find the integers.

_____

2. If the square of the sum of two consecutive integers is reduced by twice their product, the result is 25. Find the integers.

_____

Step 4 Solve.

$$x^2 + x^2 + 2x + 1 = 3 - x$$

$$2x^2 + 3x - 2 = 0$$

$$(2x - 1)(x + 2) = 0$$

$$2x + 1 = 0 \quad \text{or} \quad x + 2 = 0$$

$$x = -\frac{1}{2} \quad \text{or} \quad x = -2$$

Step 5 State the answer. The solutions are $-\frac{1}{2}$

and –2. We discard $-\frac{1}{2}$ since it is not an integer.

If $x = -2$, then $x + 1 = -2 + 1 = -1$. So, the numbers are –2 and –1.

Step 6 Check. $(-2)^2 + (-1)^2 = 3 + [-(-2)]$

$$4 + 1 = 3 + 2$$

$$5 = 5$$

Objective 2 Practice Exercises

For extra help, see Examples 2–3 on pages 474–475 of your text.

Solve each problem.

4.   Find all possible pairs of consecutive odd integers   4. _____
     whose sum is equal to their product decreased by 47.

5. Find two consecutive positive even integers whose   5. _____
    product is six more than three times its sum.

6. Find three consecutive positive odd integers such   6. _____
    that four times the sum of all three equals 13 more
    than the product of the smaller two.

Objective 3   Solve problems by applying the Pythagorean theorem.

Video Examples

Review this example for Objective 3:

4. Penny and Carla started biking from the same
    corner. Penny biked east and Carla biked south.
    When they were 26 miles apart, Carla had biked
    14 miles further than Penny. Find the distance
    each biked.

    Step 1 Read carefully. Find the two distances.

    Step 2 Assign a variable.
        Let x = Penny's distance.

Now Try:

4. A ladder is leaning against a
    building. The distance from the
    bottom of the ladder to the
    building is 8 feet less than the
    length of the ladder. How high
    up the side of the building is the
    top of the ladder if that distance
    is 4 feet less than the length of

Then $x + 14 =$ Carla's distance.

Step 3 Write an equation. Substitute into the Pythagorean theorem.

$$a^2 + b^2 = c^2$$

$$x^2 + (x+14)^2 = 26^2$$

Step 4 Solve.

$$x^2 + x^2 + 28x + 196 = 676$$

$$2x^2 + 28x - 480 = 0$$

$$2(x^2 + 14x - 240) = 0$$

$$x^2 + 14x - 240 = 0$$

$$(x+24)(x-10) = 0$$

$$x + 24 = 0 \quad \text{or} \quad x - 10 = 0$$

$$x = -24 \quad \text{or} \quad x = 10$$

Step 5 State the answer. Since –24 cannot be a distance, 10 is the distance for Penny, and 10 + 14 = 24 is the distance for Carla.

Step 6 Check. Since $10^2 + 24^2 = 26^2$ is true, the answer is correct.

the ladder?

_____

Objective 3 Practice Exercises

For extra help, see Example 4 on page 476 of your text.

Solve each problem.

7.   A field is in the shape of a right triangle. The shorter leg measures 45 meters. The hypotenuse measures 45 meters less than twice the longer the leg. Find the dimensions of the lot.

7. _____

8.   A train and a car leave a station at the same time, the   **8.** car_____
     train traveling due north and the car traveling west.
     When they are 100 miles apart, the train has traveled   train _____
     20 miles farther than the car. Find the distance each
     has traveled.

9.   Two ships left a dock at the same time. When they   **9.** _____
     were 25 miles apart, the ship that sailed due south
     had gone 10 miles less than twice the distance
     traveled by the ship that sailed due west. Find the
     distance traveled by the ship that sailed due south.

Name: _____ Date: _____
Instructor: _____ Section: _____

Objective 4   Solve problems using given quadratic models.

Video Examples

Review this example for Objective 4:

5. Jeff threw a stone straight upward at 46 feet per second from a dock 6 feet above a lake. The height of the stone above the lake t seconds after it is thrown is given by $h = -16t^2 + 46t + 6$. How long will it take for the stone to reach a height of 39 feet?

Substitute 39 for h.

$$39 = -16t^2 + 46t + 6$$

Solve for t.

$$16t^2 - 46t + 33 = 0$$

$$(8t - 11)(2t - 3) = 0$$

$$8t - 11 = 0 \quad \text{or} \quad 2t - 3 = 0$$

$$t = \frac{11}{8} \quad \text{or} \quad t = \frac{3}{2}$$

Since we have found two acceptable answers, the stone will be at height of 39 feet twice (once on its way up and once on its way down) —at $\frac{11}{8}$ sec or $\frac{3}{2}$ sec.

Now Try:

5. A ball is dropped from the roof of a 19.6 meter high building. Its height h (in meters) t seconds later is given by the equation $h = -4.9t^2 + 19.6$. After how many second is the height 14.7 meters?

_____

Objective 4 Practice Exercises

For extra help, see Examples 5–6 on pages 477–478 of your text.

Solve each problem.

10. If an object is propelled upward from a height of 16 feet with an initial velocity of 48 feet per second, its height h (in feet) t seconds later is given by the equation $h = -16t^2 + 48t + 16$.

(a) After how many seconds is the height 52 feet?

(b) After how many seconds is the height 48 feet?

10. a._____

b._____

Copyright © 2018 Pearson Education, Inc.

11. A company determines that its daily revenue R (in dollars) for selling x items is modeled by the equation $R = x(150 - x)$. How many items must be sold for its revenue to be $4400?

11. _____

12. If a ball is batted at an angle of $35°$, the distance that the ball travels is given approximately by

$D = 0.029v^2 + 0.021v - 1$, where v is the bat speed in miles per hour and D is the distance traveled in feet. Find the distance a batted ball will travel if the ball is batted with a velocity of 90 miles per hour. Round your answer to the nearest whole number.

12. _____

Chapter 7 RATIONAL EXPRESSIONS AND APPLICATIONS

7.1   The Fundamental Property of Rational Expressions

Learning Objectives
1   Find the numerical value of a rational expression.
2   Find the values of the variable for which a rational expression is undefined.
3   Write rational expressions in lowest terms.
4   Recognize equivalent forms of rational expressions.

Key Terms

Use the vocabulary terms listed below to complete each statement in exercises 1–2.

      rational expression   **lowest terms**

1.   The quotient of two polynomials with denominator not 0 is called a

    _____ .

2.   A rational expression is written in _____ if the greatest
common factor of its numerator and denominator is 1.

Objective 1   Find the numerical value of a rational expression.

Video Examples

Review these examples for Objective 1:

1.   Find the numerical value of $\dfrac{4x+8}{3x-6}$ for each
value of x.

    a. $x = 1$

$$\frac{4x+8}{3x-6} = \frac{4(1)+8}{3(1)-6} = \frac{12}{-3} = -4$$

    b. $x = 0$

$$\frac{4x+8}{3x-6} = \frac{4(0)+8}{3(0)-6} = \frac{8}{-6}, \text{ or } -\frac{4}{3}$$

    c. $x = 2$

$$\frac{4x+8}{3x-6} = \frac{4(2)+8}{3(2)-6} = \frac{16}{0} \quad \text{undefined}$$

    d. $x = -2$

$$\frac{4x+8}{3x-6} = \frac{4(-2)+8}{3(-2)-6} = \frac{0}{-12} = 0$$

Now Try:

1.   Find the numerical value of
$\dfrac{x-7}{x-5}$ for each value of x.

    a. $x = 1$

    _____

    b. $x = 0$

    _____

    c. $x = 5$

    _____

    d. $x = 7$

    _____

Objective 1 Practice Exercises

For extra help, see Example 1 on page 498 of your text.

Find the numerical value of each expression when (*a*) $x = 4$ *and* (*b*) $x = -1$.

1. $\dfrac{-3x+1}{2x+1}$

1. (a) _____

   (b) _____

2. $\dfrac{2x^2-4}{x^2-2}$

2. (a) _____

   (b) _____

3. $\dfrac{2x-5}{2-x-x^2}$

3. (a) _____

   (b) _____

Objective 2   Find the values of the variable for which a rational expression is undefined.

Video Examples

Review these examples for Objective 2:

2. Find any values of the variable for which each rational expression is undefined.

   a. $\dfrac{3x+8}{5x+4}$

   Step 1 Set the denominator equal to 0.
   $$5x+4=0$$
   Step 2 Solve.
   $$5x=-4$$
   $$x=-\frac{4}{5}$$

Now Try:

2. Find any values of the variable for which each rational expression is undefined.

   a. $\dfrac{y+6}{7y-1}$

   _____

Step 3 The given expression is undefined for $-\dfrac{4}{5}$, so $x \neq -\dfrac{4}{5}$.

b. $\dfrac{3m^2}{m^2 - 4m - 5}$

Set the denominator equal to 0.

$$m^2 - 4m - 5 = 0$$

$$(m+1)(m-5) = 0$$

$$m+1 = 0 \quad \text{or} \quad m-5 = 0$$

$$m = -1 \quad \text{or} \quad m = 5$$

The given expression is undefined for –1 and 5, so $m \neq -1$, $m \neq 5$.

c. $\dfrac{6r}{r^2 + 49}$

This denominator will not equal 0 for any value of *r*. There are no values for which this expression is undefined.

b. $\dfrac{15m^2}{m^2 - m - 20}$

_____

c. $\dfrac{12t^2}{t^2 + 100}$

_____

Objective 2 Practice Exercises

For extra help, see Example 2 on page 499 of your text.

Find any value(s) of the variable for which each rational expression is undefined. Write answers with $\neq$.

4. $\dfrac{4x^2}{x + 7}$

4. _____

5. $\dfrac{2x^2}{x^2 + 4}$

5. _____

6. $\dfrac{2y - 5}{2y^2 + 4y - 16}$

6. _____

Objective 3   Write rational expressions in lowest terms.

Video Examples

Review these examples for Objective 3: | **Now Try:**

3b. Write the expression $\dfrac{15k^3}{3k^4}$ in lowest terms.

Write k^3 as $k \cdot k \cdot k$ and k^4 as $k \cdot k \cdot k \cdot k$.

$$\frac{15k^3}{3k^4} = \frac{3 \cdot 5 \cdot k \cdot k \cdot k}{3 \cdot k \cdot k \cdot k \cdot k}$$

$$= \frac{5 \cdot (3 \cdot k \cdot k \cdot k)}{k \cdot (3 \cdot k \cdot k \cdot k)}$$

$$= \frac{5}{k}$$

3b. Write the expression $\dfrac{12k^5}{4k^8}$ in lowest terms.

_____

4. Write each rational expression in lowest terms.

a. $\dfrac{6x-18}{5x-15}$

$$\frac{6x-18}{5x-15} = \frac{6(x-3)}{5(x-3)} = \frac{6}{5}$$

c. $\dfrac{m^2+5m-24}{3m^2-5m-12}$

$$\frac{m^2+5m-24}{3m^2-5m-12} = \frac{(m+8)(m-3)}{(3m+4)(m-3)}$$

$$= \frac{m+8}{3m+4}$$

4. Write each rational expression in lowest terms.

a. $\dfrac{7x-35}{9x-45}$

_____

c. $\dfrac{m^2-3m-54}{2m^2-15m-27}$

_____

5. Write $\dfrac{3x-2y}{2y-3x}$ in lowest terms.

Factor -1 from the denominator.

$$\frac{3x-2y}{2y-3x} = \frac{3x-2y}{-1(-2y+3x)}$$

$$= \frac{3x-2y}{-1(3x-2y)}$$

$$= -1$$

Alternatively, we could factor -1 from the numerator.

$$\frac{3x-2y}{2y-3x} = \frac{-1(-3x+2y)}{2y-3x}$$

$$= \frac{-1(2y-3x)}{2y-3x}$$

$$= -1$$

5. Write $\dfrac{4y-5x}{5x-4y}$ in lowest terms.

_____

Objective 3 Practice Exercises

For extra help, see Examples 3–6 on pages 500–503 of your text.

Write each rational expression in lowest terms. Assume that no values of any variable make any denominator zero.

7. $\dfrac{15ab^3c^9}{-24ab^2c^{10}}$

7. _____

8. $\dfrac{16-x^2}{2x-8}$

8. _____

9. $\dfrac{9x^2-9x-108}{2x-8}$

9. _____

Objective 4   Recognize equivalent forms of rational expressions.

Video Examples

Review this example for Objective 4:

7. Write four equal forms of the following rational expression.

$$-\frac{4x+3}{x-8}$$

If we apply the negative sign to the numerator, we obtain the first two equivalent forms.

$$\frac{-(4x+3)}{x-8} \quad \text{and} \quad \frac{-4x-3}{x-8}$$

If we apply the negative sign to the denominator, we obtain the last two equivalent forms.

$$\frac{4x+3}{-(x-8)} \quad \text{and} \quad \frac{4x+3}{-x+8}$$

Now Try:

7. Write four equal forms of the following rational expression.

$$-\frac{10x-7}{4x-3}$$

_____

Objective 4 Practice Exercises

For extra help, see Example 7 on page 504 of your text.

Write four equivalent forms of the following rational expressions. Assume that no values of any variable make any denominator zero.

10. $-\dfrac{4x+5}{3-6x}$ **10.** _____

11. $\dfrac{2p-1}{1-4p}$ **11.** _____

12. $-\dfrac{2x-3}{x+2}$ **12.** _____

Chapter 7 RATIONAL EXPRESSIONS AND APPLICATIONS

7.2   Multiplying and Dividing Rational Expressions

Learning Objectives
1   Multiply rational expressions.
2   Find reciprocals.
3   Divide rational expressions.

Key Terms

Use the vocabulary terms listed below to complete each statement in exercises 1−3.

rational expression   **reciprocal**   **lowest terms**

1.   The _____ of the expression $\frac{4x-5}{x+2}$ is $\frac{x+2}{4x-5}$.

2.   A _____ is the quotient of two polynomials with denominator not 0.

3.   A rational expression is written in _____ when the numerator and denominator have no common terms.

Objective 1   Multiply rational expressions.

Video Examples

Review these examples for Objective 1:

1b. Multiply. Write each answer in lowest terms.

$$\frac{9}{x^2}\cdot\frac{x^3}{15}$$

$$\frac{9}{x^2}\cdot\frac{x^3}{15}=\frac{9\cdot x^3}{x^2\cdot 15}$$

$$=\frac{3\cdot 3\cdot x\cdot x\cdot x}{3\cdot 5\cdot x\cdot x}$$

$$=\frac{3x}{5}$$

2.   Multiply. Write the answer in lowest terms.

$$\frac{3x+2y}{5x}\cdot\frac{x^2}{(3x+2y)^2}$$

Multiply numerators, multiply denominators, factor, and identify the common factors.

Now Try:

1b. Multiply. Write each answer in lowest terms.

$$\frac{8}{x^3}\cdot\frac{x^2}{6}$$

_____

2.   Multiply. Write the answer in lowest terms.

$$\frac{r-s}{6s}\cdot\frac{s^3}{(r-s)^2}$$

_____

$$\frac{3x+2y}{5x} \cdot \frac{x^2}{(3x+2y)^2} = \frac{(3x+2y)x^2}{5x(3x+2y)^2}$$

$$= \frac{(3x+2y) \cdot x \cdot x}{5x(3x+2y)(3x+2y)}$$

$$= \frac{x}{5(3x+2y)}$$

Objective 1 Practice Exercises

For extra help, see Examples 1–3 on pages 508–509 of your text.

Multiply. Write each answer in lowest terms.

1. $\dfrac{8m^4n^3}{3} \cdot \dfrac{5}{4mn^2}$

1. _____

2. $\dfrac{m^2-16}{m-3} \cdot \dfrac{9-m^2}{4-m}$

2. _____

3. $\dfrac{3x+12}{6x-30} \cdot \dfrac{x^2-x-20}{x^2-16}$

3. _____

Objective 2   Find reciprocals.

Video Examples

Review these examples for Objective 2:	**Now Try:**
4. Find the reciprocal of each rational expression.	4. Find the reciprocal of each rational expression.

a. $\dfrac{7x^3}{8y}$ a. $\dfrac{11a^3}{10b}$

$\dfrac{7x^3}{8y}$ has reciprocal $\dfrac{8y}{7x^3}$.   _____

b. $\dfrac{m^2-16}{m^2-2m-15}$ b. $\dfrac{w^2-6w}{w^2+6w-7}$

$\dfrac{m^2-16}{m^2-2m-15}$ has reciprocal $\dfrac{m^2-2m-15}{m^2-16}$.   _____

Objective 2 Practice Exercises

For extra help, see Example 4 on page 509 of your text.

Write the reciprocal of each rational expression.

4. $\dfrac{7}{2y}$ 4.   _____

5. $4r^2+2r+3$ 5.   _____

6. $\dfrac{3x+4y}{5x-2y}$ 6.   _____

Objective 3   Divide rational expressions.

Video Examples

Review these examples for Objective 3:

6. Divide. Write the answer in lowest terms.

$$\frac{(5m)^2}{(3p)^3} \div \frac{10m^4}{24p^2}$$

Multiply by the reciprocal.

$$\frac{(5m)^2}{(3p)^3} \div \frac{10m^4}{24p^2} = \frac{(5m)^2}{(3p)^3} \cdot \frac{24p^2}{10m^4}$$

$$= \frac{25m^2}{27p^3} \cdot \frac{24p^2}{10m^4}$$

$$= \frac{25 \cdot 24m^2 p^2}{27 \cdot 10m^4 p^3}$$

$$= \frac{5 \cdot 5 \cdot 3 \cdot 2 \cdot 4 \cdot m^2 \cdot p^2}{3 \cdot 9 \cdot 2 \cdot 5 \cdot m^2 \cdot m^2 \cdot p^2 \cdot p}$$

$$= \frac{20}{9m^2 p}$$

7. Divide. Write the answer in lowest terms.

$$\frac{m^2 - 16}{(m-5)(m-4)} \div \frac{(m+4)(m-5)}{-6m}$$

Multiply by the reciprocal.

$$\frac{m^2 - 16}{(m-5)(m-4)} \div \frac{(m+4)(m-5)}{-6m}$$

$$= \frac{m^2 - 16}{(m-5)(m-4)} \cdot \frac{-6m}{(m+4)(m-5)}$$

$$= \frac{-6m(m^2 - 16)}{(m-5)(m-4)(m+4)(m-5)}$$

$$= \frac{-6m(m+4)(m-4)}{(m-5)(m-4)(m+4)(m-5)}$$

$$= \frac{-6m}{(m-5)^2}, \text{ or } -\frac{6m}{(m-5)^2}$$

8. Divide. Write the answer in lowest terms.

$$\frac{x^2 - 25}{x^2 - 9} \div \frac{3x^2 - 15x}{3-x}$$

Multiply by the reciprocal.

Now Try:

6. Divide. Write the answer in lowest terms.

$$\frac{(7c)^2}{(5d)^3} \div \frac{49c^2}{10d^3}$$

_____

7. Divide. Write the answer in lowest terms.

$$\frac{x^2 - 49}{(x-7)(x-3)} \div \frac{(x+7)(x-3)}{8x}$$

_____

8. Divide. Write the answer in lowest terms.

$$\frac{m^2 - 64}{m^2 - 81} \div \frac{5m^2 + 40m}{9-m}$$

_____

$$\frac{x^2-25}{x^2-9} \div \frac{3x^2-15x}{3-x}$$

$$=\frac{x^2-25}{x^2-9} \cdot \frac{3-x}{3x^2-15x}$$

$$=\frac{(x^2-25)(3-x)}{(x^2-9)(3x^2-15x)}$$

$$=\frac{(x+5)(x-5)(3-x)}{(x+3)(x-3)(3x)(x-5)}$$

$$=\frac{-1(x+5)}{3x(x+3)}, \quad \text{or} \quad \frac{-x-5}{3x(x+3)}$$

Objective 3 Practice Exercises

For extra help, see Examples 5–8 on pages 497–498 of your text and Section Lecture video for Section 7.2 and Exercise Solutions Clip 17, 19, 21, 33, and 35.

Divide. Write each answer in lowest terms.

7. $\dfrac{b-7}{16} \div \dfrac{7-b}{8}$

7. _____

8. $\dfrac{m^2+2mn+n^2}{m^2+m} \div \dfrac{m^2-n^2}{m^2-1}$

8. _____

9. $\dfrac{27-3k^2}{3k^2+8k-3} \div \dfrac{k^2-6k+9}{6k^2-19k+3}$

9. _____

Chapter 7 RATIONAL EXPRESSIONS AND APPLICATIONS

7.3   Least Common Denominators

Learning Objectives
1   Find the least common denominator for a list of fractions.
2   Write equivalent rational expressions.

Key Terms

Use the vocabulary terms listed below to complete each statement in exercises 1−2.

least common denominator   **equivalent expressions**

1. $\dfrac{24x-8}{9x^2-1}$ and $\dfrac{8}{3x+1}$ are _____.

2.   The simplest expression that is divisible by all denominators is called the

_____.

Objective 1   Find the least common denominator for a list of fractions.

Video Examples

Review these examples for Objective 1:

2.   Find the LCD for $\dfrac{9}{28s^3}$ and $\dfrac{5}{42s^2}$.

Step 1
$$28s^3 = 2 \cdot 2 \cdot 7 \cdot s^3$$
$$42s^2 = 2 \cdot 3 \cdot 7 \cdot s^2$$

Step 2
Here s appears three times, 2 appears twice, and 3 and 7 each appear once.

Step 3
$$\text{LCD} = 2^2 \cdot 3 \cdot 7 \cdot s^3 = 84s^3$$

3.   Find the LCD for the fractions in each list.

a. $\dfrac{5}{7b}, \dfrac{8}{b^2-5b}$

$$7b = 7 \cdot b$$
$$b^2 - 5b = b(b-5)$$
$$\text{LCD} = 7 \cdot b(b-5) = 7b(b-5)$$

Now Try:

2.   Find the LCD for

$\dfrac{17}{40a^2}$ and $\dfrac{13}{24a^4}$.

_____

3.   Find the LCD for the fractions in each list.

a. $\dfrac{7}{9w}, \dfrac{13}{w^2-2w}$

_____

b. $\dfrac{6}{c^2-5c-6}, \dfrac{10}{c^2+3c-54}, \dfrac{4}{c^2-12c+36}$

$$c^2-5c-6=(c-6)(c+1)$$
$$c^2+3c-54=(c-6)(c+9)$$
$$c^2-12c+36=(c-6)^2$$

Use each factor the greatest number of times it appears as a factor.

$$\text{LCD}=(c+1)(c+9)(c-6)^2$$

b. $\dfrac{12}{b^2-16}, \dfrac{6}{b^2-3b-4},$

$$\dfrac{9}{b^2-8b+16}$$

_____

Objective 1 Practice Exercises

For extra help, see Examples 1–3 on pages 514–515 of your text.

Find the least common denominator for each list of rational expressions.

1. $\dfrac{13}{36b^4}, \dfrac{17}{27b^2}$

1. _____

2. $\dfrac{-7}{a^2-2a}, \dfrac{3a}{2a^2+a-10}$

2. _____

3. $\dfrac{8}{w^3-9w}, \dfrac{4w}{w^2+w-6}$

3. _____

Objective 2   Write equivalent rational expressions.

Video Examples

Review these examples for Objective 2:

4a. Write the rational expression as an equivalent expression with the indicated denominator.

$$\frac{7}{9} = \frac{?}{45}$$

Step 1 Factor both denominators.

$$\frac{7}{9} = \frac{?}{5 \cdot 9}$$

Step 2 A factor of 5 is missing.

Step 3 Multiply $\frac{7}{9}$ by $\frac{5}{5}$.

$$\frac{7}{9} = \frac{7}{9} \cdot \frac{5}{5} = \frac{35}{45}$$

5. Write each rational expression as an equivalent expression with the indicated denominator.

a. $\dfrac{9}{5x+2} = \dfrac{?}{15x+6}$

Factor the denominator on the right.

$$\frac{9}{5x+2} = \frac{?}{3(5x+2)}$$

The missing factor is 3, so multiply by $\frac{3}{3}$.

$$\frac{9}{5x+2} \cdot \frac{3}{3} = \frac{27}{15x+6}$$

b. $\dfrac{7}{q^2+6q} = \dfrac{?}{q^3+q^2-30q}$

Factor the denominator in each rational expression.

$$\frac{7}{q(q+6)} = \frac{?}{q(q+6)(q-5)}$$

The missing factor is $(q-5)$, so multiply by $\dfrac{q-5}{q-5}$.

$$\frac{7}{q(q+6)} = \frac{7}{q(q+6)} \cdot \frac{q-5}{q-5}$$

$$= \frac{7(q-5)}{q^3-q^2-30q}$$

$$= \frac{7q-35}{q^3-q^2-30q}$$

Now Try:

4a. Write the rational expression as an equivalent expression with the indicated denominator.

$$\frac{13}{6} = \frac{?}{30}$$

_____

5. Write each rational expression as an equivalent expression with the indicated denominator.

a. $\dfrac{19}{6c-5} = \dfrac{?}{24c-20}$

_____

b. $\dfrac{3}{z^2-7z} = \dfrac{?}{z^3-5z^2-14z}$

_____

Objective 2 Practice Exercises

For extra help, see Examples 4–5 on pages 516–517 of your text.

Rewrite each rational expression with the indicated denominator. Give the numerator of the new fraction.

4. $\dfrac{5a}{8a-3} = \dfrac{?}{6-16a}$

4. _____

5. $\dfrac{3}{5r-10} = \dfrac{?}{50r^2-100r}$

5. _____

6. $\dfrac{3}{k^2+3k} = \dfrac{?}{k^3+10k^2+21k}$

6. _____

Chapter 7 RATIONAL EXPRESSIONS AND APPLICATIONS

7.4   Adding and Subtracting Rational Expressions

Learning Objectives
1   Add rational expressions having the same denominator.
2   Add rational expressions having different denominators.
3   Subtract rational expressions.

Key Terms

Use the vocabulary terms listed below to complete each statement in exercises 1–2.

least common multiple   greatest common factor

1.   The _____ of $2m^2 - 5m - 3$ and $2m - 6$ is $m - 3$.

2.   The _____ of $2m^2 - 5m - 3$ and $2m - 6$ is
     $2(m-3)(2m+1)$.

Objective 1   Add rational expressions having the same denominator.

Video Examples

Review this example for Objective 1:
1b. Add. Write each answer in lowest terms.

$$\frac{4x}{x+2} + \frac{8}{x+2}$$

$$\frac{4x}{x+2} + \frac{8}{x+2} = \frac{4x+8}{x+2} = \frac{4(x+2)}{x+2} = 4$$

Now Try:
1b. Add. Write each answer in lowest terms.

$$\frac{2x^2}{x+4} + \frac{8x}{x+4}$$

_____

Objective 1 Practice Exercises

For extra help, see Example 1 on page 520 of your text.

Add. Write each answer in lowest terms.

1. $\dfrac{5}{3w^2} + \dfrac{7}{3w^2}$

1.   _____

2. $\dfrac{b}{b^2 - 4} + \dfrac{2}{b^2 - 4}$

2.   _____

3. $\dfrac{2x+3}{x^2+3x-10} + \dfrac{2-x}{x^2+3x-10}$

3. _____

Objective 2   Add rational expressions having different denominators.

Video Examples

Review these examples for Objective 2:

2a. Add. Write each answer in lowest terms.

$$\frac{5}{18} + \frac{7}{24}$$

Step 1 Find the LCD.

$$18 = 2 \cdot 3 \cdot 3 = 2 \cdot 3^2$$

$$24 = 2 \cdot 2 \cdot 2 \cdot 3 = 2^3 \cdot 3$$

$$\text{LCD} = 2^3 \cdot 3^2 = 72$$

Step 2 Now write each expression as an equivalent expression with the LCD as the denominator.

$$\frac{5}{18} + \frac{7}{24} = \frac{5(4)}{18(4)} + \frac{7(3)}{24(3)}$$

$$= \frac{20}{72} + \frac{21}{72}$$

Step 3 Add the numerators.

Step 4 Write in lowest terms, if necessary.

$$= \frac{20+21}{72}$$

$$= \frac{41}{72}$$

3. Add. Write the answer in lowest terms.

$$\frac{10x}{x^2-9} + \frac{-5}{x+3}$$

Step 1 Find the LCD.

$$x^2-9 = (x+3)(x-3)$$

$$x+3 \text{ is prime}$$

$$\text{LCD} = (x+3)(x-3)$$

Step 2 Now write each expression as an equivalent expression with the LCD as the

Now Try:

2a. Add. Write each answer in lowest terms.

$$\frac{9}{35} + \frac{8}{45}$$

_____

3. Add. Write the answer in lowest terms.

$$\frac{8x}{x^2-100} + \frac{-4}{x+10}$$

_____

denominator.

$$\frac{10x}{x^2-9}+\frac{-5}{x+3}$$

$$=\frac{10x}{(x+3)(x-3)}+\frac{-5(x-3)}{(x+3)(x-3)}$$

$$=\frac{10x}{(x+3)(x-3)}+\frac{-5x+15}{(x+3)(x-3)}$$

Step 3 Add the numerators.

$$=\frac{10x-5x+15}{(x+3)(x-3)}$$

$$=\frac{5x+15}{(x+3)(x-3)}$$

Step 4 Write in lowest terms, if necessary.

$$=\frac{5(x+3)}{(x+3)(x-3)}$$

$$=\frac{5}{x-3}$$

4. Add. Write the answer in lowest terms.

$$\frac{3x}{x^2-x-20}+\frac{5}{x^2-2x-15}$$

The LCD is $(x+3)(x+4)(x-5)$.

$$=\frac{3x}{(x+4)(x-5)}+\frac{5}{(x+3)(x-5)}$$

$$=\frac{3x(x+3)}{(x+3)(x+4)(x-5)}+\frac{5(x+4)}{(x+3)(x+4)(x-5)}$$

$$=\frac{3x(x+3)+5(x+4)}{(x+3)(x+4)(x-5)}$$

$$=\frac{3x^2+9x+5x+20}{(x+3)(x+4)(x-5)}$$

$$=\frac{3x^2+14x+20}{(x+3)(x+4)(x-5)}$$

4. Add. Write the answer in lowest terms.

$$\frac{7x}{x^2-2x-8}+\frac{4}{x^2-3x-4}$$

_____

Objective 2 Practice Exercises

For extra help, see Examples 2–5 on pages 521–523 of your text.

Add. Write each answer in lowest terms.

4. $\dfrac{7}{x-5}+\dfrac{4}{x+5}$

4. _____

5. $\dfrac{3z}{z^2-4}+\dfrac{4z-3}{z^2-4z+4}$

5. _____

6. $\dfrac{4z}{z^2+6z+8}+\dfrac{2z-1}{z^2+5z+6}$

6. _____

Objective 3   Subtract rational expressions.

Video Examples

Review these examples for Objective 3:
8. Subtract. Write the answer in lowest terms.

$\dfrac{5x}{x-7}-\dfrac{x-42}{7-x}$

The denominators are opposites.

$\dfrac{5x}{x-7}-\dfrac{x-42}{7-x}=\dfrac{5x}{x-7}-\dfrac{(x-42)(-1)}{(7-x)(-1)}$

$=\dfrac{5x}{x-7}-\dfrac{-x+42}{x-7}$

$=\dfrac{5x+x-42}{x-7}$

$=\dfrac{6x-42}{x-7}$

$=\dfrac{6(x-7)}{x-7}$

$=6$

Now Try:
8. Subtract. Write the answer in lowest terms.

$\dfrac{4x}{x-9}-\dfrac{3x-63}{9-x}$

_____

9. Subtract. Write the answer in lowest terms.

$$\frac{7x}{x^2-4x+4}-\frac{2}{x^2-4}$$

$$=\frac{7x}{(x-2)^2}-\frac{2}{(x+2)(x-2)}$$

The LCD is $(x+2)(x-2)^2$.

$$=\frac{7x}{(x-2)^2}-\frac{2}{(x+2)(x-2)}$$

$$=\frac{7x(x+2)}{(x+2)(x-2)^2}-\frac{2(x-2)}{(x+2)(x-2)^2}$$

$$=\frac{7x(x+2)-2(x-2)}{(x+2)(x-2)^2}$$

$$=\frac{7x^2+14x-2x+4}{(x+2)(x-2)^2}$$

$$=\frac{7x^2+12x+4}{(x+2)(x-2)^2}$$

10. Subtract. Write the answer in lowest terms.

$$\frac{2q}{q^2-2q-8}-\frac{7}{3q^2-14q+8}$$

The LCD is $(q+2)(3q-2)(q-4)$.

$$=\frac{2q}{(q+2)(q-4)}-\frac{7}{(3q-2)(q-4)}$$

$$=\frac{2q(3q-2)}{(q+2)(3q-2)(q-4)}-\frac{7(q+2)}{(q+2)(3q-2)(q-4)}$$

$$=\frac{6q^2-4q-7q-14}{(q+2)(3q-2)(q-4)}$$

$$=\frac{6q^2-11q-14}{(q+2)(3q-2)(q-4)}$$

9. Subtract. Write the answer in lowest terms.

$$\frac{8x}{x^2-10x+25}-\frac{3}{x^2-25}$$

_____

10. Subtract. Write the answer in lowest terms.

$$\frac{5}{p^2-6p+5}-\frac{4}{p^2-1}$$

_____

Name: Date:

Instructor: Section:

Objective 3 Practice Exercises

For extra help, see Examples 6–10 on pages 523–525 of your text.

Subtract. Write each answer in lowest terms.

7. $\dfrac{z+2}{z-2} - \dfrac{z-2}{z+2}$

                                                     7. _____

8. $\dfrac{-4}{x^2-4} - \dfrac{3}{4-2x}$

                                                     8. _____

9. $\dfrac{m}{m^2-4} - \dfrac{1-m}{m^2+4m+4}$

                                                     9. _____

Chapter 7 RATIONAL EXPRESSIONS AND APPLICATIONS

7.5   Complex Fractions

Learning Objectives
1   Define and recognize a complex fraction.
2   Simplify a complex fraction by writing it as a division problem (Method 1).
3   Simplify a complex fraction by multiplying numerator and denominator by the least common denominator (Method 2).

Key Terms

Use the vocabulary terms listed below to complete each statement in exercises 1−2.

complex fraction   **LCD**

1.   A _____ is a rational expression with one or more fractions
       in the numerator, denominator, or both.

2.   To simplify a complex fraction, multiply the numerator and denominator by the
       _____ of all the fractions within the complex fraction.

Objective 1   Define and recognize a complex fraction.

For extra help, see page 530 of your text.

**Objective 2   Simplify a complex fraction by writing it as a division problem
                 (Method 1).**

Video Examples

Review these examples for Objective 2:

1b. Simplify the complex fraction.

$$\frac{8+\dfrac{4}{x}}{\dfrac{x}{6}+\dfrac{1}{12}}$$

Step 1 Write the numerator as a single fraction.

$$8+\frac{4}{x}=\frac{8}{1}+\frac{4}{x}=\frac{8x}{x}+\frac{4}{x}=\frac{8x+4}{x}$$

Do the same with each denominator.

$$\frac{x}{6}+\frac{1}{12}=\frac{x(2)}{6(2)}+\frac{1}{12}=\frac{2x}{12}+\frac{1}{12}=\frac{2x+1}{12}$$

Step 2 Write the equivalent complex fraction as
a division problem.

$$\frac{\dfrac{8x+4}{x}}{\dfrac{2x+1}{12}}=\frac{8x+4}{x}\div\frac{2x+1}{12}$$

Now Try:

1b. Simplify the complex fraction.

$$\frac{9+\dfrac{3}{x}}{\dfrac{x}{10}+\dfrac{1}{30}}$$

_____

Step 3 Divide by multiplying by the reciprocal.

$$\frac{8x+4}{x} \div \frac{2x+1}{12} = \frac{8x+4}{x} \cdot \frac{12}{2x+1}$$

$$= \frac{4(2x+1)}{x} \cdot \frac{12}{2x+1}$$

$$= \frac{48}{x}$$

2. Simplify the complex fraction.

$$\frac{\dfrac{rs^2}{t^3}}{\dfrac{s^3}{r^3t}}$$

Use the definition of division and then the fundamental property.

$$= \frac{rs^2}{t^3} \div \frac{s^3}{r^3t}$$

$$= \frac{rs^2}{t^3} \cdot \frac{r^3t}{s^3}$$

$$= \frac{r^4}{st^2}$$

3. Simplify the complex fraction.

$$\frac{\dfrac{30}{x+4} - 6}{\dfrac{4}{x+4} + 1} = \frac{\dfrac{30}{x+4} - \dfrac{6(x+4)}{x+4}}{\dfrac{4}{x+4} + \dfrac{1(x+4)}{x+4}}$$

$$= \frac{\dfrac{30 - 6(x+4)}{x+4}}{\dfrac{4 + 1(x+4)}{x+4}}$$

$$= \frac{\dfrac{30 - 6x - 24}{x+4}}{\dfrac{4 + x + 4}{x+4}}$$

$$= \frac{\dfrac{6 - 6x}{x+4}}{\dfrac{x+8}{x+4}}$$

$$= \frac{6 - 6x}{x+4} \cdot \frac{x+4}{x+8}$$

$$= \frac{6 - 6x}{x+8}$$

2. Simplify the complex fraction.

$$\frac{\dfrac{a^3b^2}{c}}{\dfrac{a^5b}{c^3}}$$

_____

3. Simplify the complex fraction.

$$\frac{\dfrac{20}{x-5} - 9}{\dfrac{5}{x-5} + 2}$$

_____

Name: _____ Date: _____

Instructor: _____ Section: _____

Objective 2 Practice Exercises

For extra help, see Examples 1–3 on pages 531–532 of your text.

Simplify each complex fraction by writing it as a division problem.

1. $\dfrac{\dfrac{49m^3}{18n^5}}{\dfrac{21m}{27n^2}}$

1. _____

2. $\dfrac{\dfrac{p}{2}-\dfrac{1}{3}}{\dfrac{p}{3}+\dfrac{1}{6}}$

2. _____

3. $\dfrac{3+\dfrac{4}{s}}{2s+\dfrac{2}{3}}$

3. _____

Objective 3 Simplify a complex fraction by multiplying numerator and denominator by the least common denominator (Method 2).

Video Examples

Review these examples for Objective 3:

4b. Simplify the complex fraction.

$$\dfrac{12+\dfrac{4}{x}}{\dfrac{x}{5}+\dfrac{1}{15}}$$

Step 1 Find the LCD for all the denominators.

The LCD for x, 5, and 15 is $15x$.

Step 2 Multiply the numerator and denominator of the complex fraction by the LCD.

$$\dfrac{12+\dfrac{4}{x}}{\dfrac{x}{5}+\dfrac{1}{15}}=\dfrac{15x\left(12+\dfrac{4}{x}\right)}{15x\left(\dfrac{x}{5}+\dfrac{1}{15}\right)}$$

$$=\dfrac{180x+60}{3x^2+x}$$

$$=\dfrac{60(3x+1)}{x(3x+1)}$$

$$=\dfrac{60}{x}$$

5. Simplify the complex fraction.

$$\dfrac{\dfrac{9}{7n}-\dfrac{3}{n^2}}{\dfrac{8}{3n}+\dfrac{5}{6n^2}}$$

The LCD is $42n^2$.

$$=\dfrac{42n^2\left(\dfrac{9}{7n}-\dfrac{3}{n^2}\right)}{42n^2\left(\dfrac{8}{3n}+\dfrac{5}{6n^2}\right)}$$

$$=\dfrac{42n^2\left(\dfrac{9}{7n}\right)-42n^2\left(\dfrac{3}{n^2}\right)}{42n^2\left(\dfrac{8}{3n}\right)+42n^2\left(\dfrac{5}{6n^2}\right)}$$

$$=\dfrac{54n-126}{112n+35},\text{ or }\dfrac{18(3n-7)}{7(16n+5)}$$

Now Try:

4b. Simplify the complex fraction.

$$\dfrac{4+\dfrac{2}{x}}{\dfrac{x}{3}+\dfrac{1}{6}}$$

_____

5. Simplify the complex fraction.

$$\dfrac{\dfrac{2}{9n}-\dfrac{2}{5n^2}}{\dfrac{4}{5n}+\dfrac{2}{3n^2}}$$

_____

Name: Date:

Instructor: Section:

Objective 3 Practice Exercises

For extra help, see Examples 4–6 on pages 533–535 of your text.

Simplify each complex fraction by multiplying numerator and denominator by the least common denominator.

4. $\dfrac{\dfrac{9}{x^2}-1}{\dfrac{3}{x}-1}$

4. _____

5. $\dfrac{\dfrac{x-2}{x+2}}{\dfrac{x}{x-2}}$

5. _____

6. $\dfrac{\dfrac{6}{k+1}-\dfrac{5}{k-3}}{\dfrac{3}{k-3}+\dfrac{2}{k+2}}$

6. _____

Chapter 7 RATIONAL EXPRESSIONS AND APPLICATIONS

7.6   Solving Equations with Rational Expressions

Learning Objectives
1   Distinguish between operations with rational expressions and equations with terms that are rational expressions.
2   Solve equations with rational expressions.
3   Solve a formula for a specified variable.

Key Terms

Use the vocabulary terms listed below to complete each statement in exercises 1–2.

proposed solution   extraneous solution

1.   A solution that is not an actual solution of a given equation is called a(n)

_____ .

2.   A value of the variable that appears to be a solution after both sides of an equation with rational expressions are multiplied by a variable expression is called a(n)

_____ .

Objective 1   Distinguish between operations with rational expressions and equations with terms that are rational expressions.

Video Examples

Review these examples for Objective 1:

1.   Identify each of the following as an expression or an equation. Then simplify the expression or solve the equation.

   a. $\dfrac{8}{9}x - \dfrac{5}{6}x$

   This is a difference of two terms. It represents an expression since there is no equality symbol.
   The LCD is 18.

   $$\frac{8}{9}x - \frac{5}{6}x = \frac{2\cdot 8}{2\cdot 9}x - \frac{3\cdot 5}{3\cdot 6}x$$

   $$= \frac{16}{18}x - \frac{15}{18}x$$

   $$= \frac{1}{18}x$$

Now Try:

1.   Identify each of the following as an expression or an equation. Then simplify the expression or solve the equation.

   a. $\dfrac{4}{5}x - \dfrac{3}{10}x$

   _____

b. $\dfrac{8}{9}x - \dfrac{5}{6}x = \dfrac{2}{3}$

Because there is an equality symbol, this is an equation to be solved. The LCD is 18.

$$18\left(\frac{8}{9}x - \frac{5}{6}x\right) = 18\left(\frac{2}{3}\right)$$

$$18\left(\frac{8}{9}x\right) - 18\left(\frac{5}{6}x\right) = 18\left(\frac{2}{3}\right)$$

$$16x - 15x = 12$$

$$x = 12$$

Check
$$\frac{8}{9}x - \frac{5}{6}x = \frac{2}{3}$$

$$\frac{8}{9}(12) - \frac{5}{6}(12) \overset{?}{=} \frac{2}{3}$$

$$\frac{32}{3} - 10 \overset{?}{=} \frac{2}{3}$$

$$\frac{2}{3} = \frac{2}{3} \quad \text{True}$$

The solution set is {12}.

b. $\dfrac{4}{5}x - \dfrac{3}{10}x = 7$

_____

Objective 1 Practice Exercises

For extra help, see Example 1 on page 538 of your text.

Identify each of the following as an expression or an equation. Then simplify the expression or solve the equation.

1. $\dfrac{3x}{5} - \dfrac{4x}{3} = \dfrac{22}{15}$

1. _____

2. $\dfrac{4x}{5} - \dfrac{5x}{10}$

2. _____

3. $\dfrac{2x}{5} + \dfrac{7x}{3}$

Objective 2   Solve equations with rational expressions.

Video Examples

Review these examples for Objective 2:

7. Solve, and check the proposed solution(s).

$$\frac{6}{x^2-1} = 1 - \frac{3}{x+1}$$

$x \neq 1, -1$ or a denominator is 0.

Factor the denominator.

$$\frac{6}{(x+1)(x-1)} = 1 - \frac{3}{x+1}$$

The LCD is $(x+1)(x-1)$.

$$(x+1)(x-1)\frac{6}{(x+1)(x-1)}$$

$$= (x+1)(x-1)\left(1 - \frac{3}{x+1}\right)$$

$$(x+1)(x-1)\frac{6}{(x+1)(x-1)}$$

$$= (x+1)(x-1) - (x+1)(x-1)\frac{3}{x+1}$$

$$6 = (x+1)(x-1) - 3(x-1)$$

$$6 = x^2 - 1 - 3x + 3$$

$$0 = x^2 - 3x - 4$$

$$0 = (x+1)(x-4)$$

$$x+1 = 0 \quad \text{or} \quad x-4 = 0$$

$$x = -1 \quad \text{or} \qquad x = 4$$

Since -1 makes the original denominator equal 0, the proposed solution -1 is an extraneous value.

A check shows that $\{4\}$ is the solution set.

Now Try:

7. Solve, and check the proposed solution(s).

$$\frac{x}{x+1} + \frac{4}{x} = \frac{4}{x^2+x}$$

_____

6. Solve, and check the proposed solution.

$$\frac{2x}{x^2-25}+\frac{5}{x-5}=\frac{3}{x+5}$$

$x\neq 5,-5$ or a denominator is 0.
Factor the denominator.

$$\frac{2x}{(x+5)(x-5)}+\frac{5}{x-5}=\frac{3}{x+5}$$

The LCD is $(x+5)(x-5)$.

$$(x+5)(x-5)\left(\frac{2x}{(x+5)(x-5)}+\frac{5}{x-5}\right)$$

$$=(x+5)(x-5)\frac{3}{x+5}$$

$$(x+5)(x-5)\frac{2x}{(x+5)(x-5)}+(x+5)(x-5)\frac{5}{x-5}$$

$$=(x+5)(x-5)\frac{3}{x+5}$$

$$2x+5(x+5)=3(x-5)$$
$$2x+5x+25=3x-15$$
$$7x+25=3x-15$$
$$4x=-40$$
$$x=-10$$

A check will verify that $\{-10\}$ is the solution set.

5. Solve, and check the proposed solution.

$$\frac{3}{x^2-4}=\frac{5}{x^2-2x}$$

Step 1 Factor the denominators to find the LCD, $x(x+2)(x-2)$. Multiply by the LCD.

$$\frac{3}{(x+2)(x-2)}=\frac{5}{x(x-2)}$$

$$x(x+2)(x-2)\frac{3}{(x+2)(x-2)}$$

$$=x(x+2)(x-2)\frac{5}{x(x-2)}$$

Step 2 Solve for x.

$$3x=5(x+2)$$
$$3x=5x+10$$
$$-2x=10$$
$$x=-5$$

6. Solve, and check the proposed solution.

$$\frac{8r}{9r^2-1}=\frac{4}{3r+1}+\frac{4}{3r-1}$$

_____

5. Solve, and check the proposed solution.

$$\frac{4}{x^2-3x}=\frac{6}{x^2-9}$$

_____

Name: Date:
Instructor: Section:

Step 3 Check the proposed solution.

$$\frac{3}{x^2-4}=\frac{5}{x^2-2x}$$

$$\frac{3}{(-5)^2-4}\overset{?}{=}\frac{5}{(-5)^2-2(-5)}$$

$$\frac{3}{25-4}\overset{?}{=}\frac{5}{25+10}$$

$$\frac{1}{7}=\frac{1}{7}\quad\text{True}$$

The solution set is $\{-5\}$.

Objective 2 Practice Exercises

For extra help, see Examples 2–8 on pages 539–544 of your text.

Solve each equation and check your solutions.

4. $\dfrac{4}{n+2}-\dfrac{2}{n}=\dfrac{1}{6}$

4. _____

5. $\dfrac{x}{3x+16}=\dfrac{4}{x}$

5. _____

6. $\dfrac{-16}{n^2 - 8n + 12} = \dfrac{3}{n-2} + \dfrac{n}{n-6}$

6. _____

Objective 3   Solve a formula for a specified variable.

Video Examples

Review this example for Objective 3:

9.   Solve the formula for the specified variable.

$r = \dfrac{s+w}{v}$ for w

Isolate w. Multiply by v.

$$r = \dfrac{s+w}{v}$$

$$rv = s+w$$

$$rv - s = w$$

Check $r = \dfrac{s+w}{v}$

$$r \overset{?}{=} \dfrac{s + rv - s}{v}$$

$$r \overset{?}{=} \dfrac{rv}{v}$$

$$r = r \quad \text{True}$$

Now Try:

9.   Solve the formula for the specified variable.

$a = \dfrac{b-c}{q}$ for b

_____

Objective 3 Practice Exercises

For extra help, see Examples 9–11 on pages 544–545 of your text.

Solve each formula for the specified variable.

7. $\dfrac{1}{f} = \dfrac{1}{d_0} + \dfrac{1}{d_1}$ for f **7.** _____

8. $m = \dfrac{y_2 - y_1}{x_2 - x_1}$ for y_1 **8.** _____

9. $A = \dfrac{2pf}{b(q+1)}$ for q **9.** _____

Chapter 7 RATIONAL EXPRESSIONS AND APPLICATIONS

7.7   Applications of Rational Expressions

Learning Objectives
1   Solve problems about numbers.
2   Solve problems about distance, rate, and time.
3   Solve problems about work.

Key Terms

Use the vocabulary terms listed below to complete each statement in exercises 1–3.

reciprocal   numerator   denominator

1.   In the fraction $\frac{x+5}{x-2}$, $x+5$ is the _____.

2.   In the fraction $\frac{x+5}{x-2}$, $x-2$ is the _____.

3.   The fraction $\frac{x+5}{x-2}$ is the _____ of the fraction $\frac{x-2}{x+5}$.

Objective 1   Solve problems about numbers.

Video Examples

Review this example for Objective 1:

1.   If a certain number is added to the numerator and twice that number is subtracted from the denominator of the fraction $\frac{3}{5}$, the result is equal to 5. Find the number.

Step 1 Read the problem carefully. We are trying to find a number.

Step 2 Assign a variable.
    Let x = the number.

Step 3 Write an equation. The fraction $\frac{3+x}{5-2x}$ represents adding the number to the numerator and twice that number is subtracted from the denominator of the fraction $\frac{3}{5}$. The result is equal to 5.
$$\frac{3+x}{5-2x} = 5$$

Now Try:

1.   If the same number is added to the numerator and denominator of the fraction $\frac{5}{9}$, the value of the resulting fraction is $\frac{2}{3}$. Find the number.

    _____

Step 4 Solve. Multiply by the LCD, $5 - 2x$.

$$(5-2x)\frac{3+x}{5-2x} = (5-2x)5$$

$$3 + x = 25 - 10x$$

$$11x = 22$$

$$x = 2$$

Step 5 State the answer. The number is 2.

Step 6 Check the solution in the original problem. If 2 is added to the numerator, and twice 2 is subtracted from the denominator of $\frac{3}{5}$,

the result is $\frac{3+2}{5-2(2)} = \frac{5}{1}$, or 5, as required.

Objective 1 Practice Exercises

For extra help, see Example 1 on page 552 of your text.

Solve each problem. Check your answers to be sure they are reasonable.

1.   If three times a number is subtracted from twice its   1. _____
     reciprocal, the result is –1. Find the number.

2.   If two times a number is added to one-half of its   2. _____
     reciprocal, the result is $\frac{13}{6}$. Find the number.

3. The denominator of a fraction is 1 less than twice the   3. _____
    numerator. If the numerator and the denominator are
    each increased by 3, the resulting fraction simplifies
    to $\frac{3}{4}$. Find the original fraction.

Objective 2   Solve problems about distance, rate, and time.

Video Examples

Review this example for Objective 2:

4. A boat goes 6 miles per hour in still water. It
    takes as long to go 40 miles upstream as 80
    miles downstream. Find the speed of the current.

    Step 1 Read the problem carefully. Find the
    speed of the current.

    Step 2 Assign a variable.
    Let x = the speed of the current.
    The rate of traveling upstream is $6 - x$.
    The rate of traveling downstream is $6 + x$.

	d	r	t
Upstream	40	$6-x$	$\dfrac{40}{6-x}$
Downstream	80	$6+x$	$\dfrac{80}{6+x}$

    The times are equal.

    Step 3 Write an equation.
    $$\frac{40}{6-x} = \frac{80}{6+x}$$

Now Try:

4. The Cuyahoga River has a
    current of 2 miles per hour. Ali
    can paddle 10 miles downstream
    in the time it takes her to paddle
    2 miles upstream. How fast can
    Ali paddle?

    _____

Step 4 Solve. The LCD is $(6+x)(6-x)$.

$$(6+x)(6-x)\frac{40}{6-x} = (6+x)(6-x)\frac{80}{6+x}$$

$$40(6+x) = 80(6-x)$$

$$240+40x = 480-80x$$

$$240+120x = 480$$

$$120x = 240$$

$$x = 2$$

Step 5 State the answer. The speed of the current is 2 miles per hour.

Step 6 Check.

Upstream: $\dfrac{40}{6-2} = 10$ hr

Downstream: $\dfrac{80}{6+2} = 10$ hr

The time upstream is the same as the time downstream, as required.

Objective 2 Practice Exercises

For extra help, see Examples 2–4 on pages 553–556 of your text.

Solve each problem.

4. A boat travels 15 miles per hour in still water. The boat travels 20 miles downstream in the same time it takes the boat to travel 10 miles upstream. How fast is the current?

4. _____

5. A ship goes 120 miles downriver in $2\frac{2}{3}$ hours less than it takes to go the same distance upriver. If the speed of the current is 6 miles per hour, find the speed of the ship.

5. _____

6. On Saturday, Pablo jogged 6 miles. On Monday, jogging at the same speed, it took him 30 minutes longer to cover 10 miles. How fast did Pablo jog?

6. _____

Objective 3   Solve problems about work.

Video Examples

Review this example for Objective 3:

5. One pipe can fill a swimming pool in 8 hours and another pipe can fill the pool in 12 hours. How long will it take to fill the pool if both pipes are open?

Step 1 Read the problem carefully. Find the time working together.

Step 2 Assign a variable.
Let x = the number of hours working together.

The rate for the first pipe is $\frac{1}{8}$.

The rate for the second pipe is $\frac{1}{12}$.

Now Try:

5. Chuck can weed the garden in $\frac{1}{2}$ hour, but David takes 2 hours. How long does it take them to weed the garden if they work together?

_____

Step 3 Write an equation. The sum of the fractional part for each pipe multiplied by the time working together is the whole job.

$$\frac{1}{8}x + \frac{1}{12}x = 1$$

Step 4 Solve. The LCD is 48.

$$48\left(\frac{1}{8}x + \frac{1}{12}x\right) = 48(1)$$

$$48\left(\frac{1}{8}x\right) + 48\left(\frac{1}{12}x\right) = 48$$

$$6x + 4x = 48$$

$$10x = 48$$

$$x = \frac{48}{10} = \frac{24}{5} = 4\frac{4}{5}$$

Step 5 State the answer. Working together, it takes $4\frac{4}{5}$ hours to fill the pool.

Step 6 Check. Substitute $\frac{24}{5}$ for x in the equation from Step 3.

$$\frac{1}{8}x + \frac{1}{12}x = 1$$

$$\frac{1}{8}\left(\frac{24}{5}\right) + \frac{1}{12}\left(\frac{24}{5}\right) = 1$$

$$\frac{3}{5} + \frac{2}{5} = 1 \quad \text{True}$$

Objective 3 Practice Exercises

For extra help, see Example 5 on page 557 of your text.

Solve each problem.

7. Kelly can clean the house in 6 hours, but it takes Linda 4 hours. How long would it take them to clean the house if they worked together?

7. _____

8. Michael can type twice as fast as Sharon. Together   8. _____
    they can type a certain job in 2 hours. How long
    would it take Michael to type the entire job by
    himself?

9. A swimming pool can be filled by an inlet pipe in 18   9. _____
    hours and emptied by an outlet pipe in 24 hours.
    How long will it take to fill the empty pool if the
    outlet pipe is accidentally left open at the same time
    as the inlet pipe is opened?

Chapter 7 RATIONAL EXPRESSIONS AND APPLICATIONS

7.8   Variation

Learning Objectives
1   Solve direct variation problems.
2   Solve inverse variation problems.

Key Terms

Use the vocabulary terms listed below to complete each statement in exercises 1−3.

direct variation   **constant of variation**   **inverse variation**

1. In the equation $y = kx$, the number k is called the _____.

2. If two positive quantities x and y are in _____ and the constant of variation is positive, then as x increases, y also increases.

3. If two positive quantities x and y are in _____ and the constant of variation is positive, then as x increases, y decreases.

Objective 1   Solve direct variation problems.

Video Examples

Review these examples for Objective 1:

1. If w varies directly as v, and $w = 24$ when $v = 20$, find w when $v = 25$.

   Since w varies directly as v, there is a constant k such that $w = kv$. We let $w = 24$, $v = 20$, and solve for k.

   $$w = kv$$
   $$24 = k \cdot 20$$
   $$\frac{6}{5} = k$$

   Since $w = kv$ and $k = \frac{6}{5}$, we have $w = \frac{6}{5}v$.

   Now we can find the value of w when $v = 25$.

   $$w = \frac{6}{5} \cdot 25 = 30$$

   Thus, $w = 30$ when $v = 25$.

Now Try:

1. If a varies directly as b, and $a = 61.5$ when $b = 82$, find a when $b = 224$.

   _____

2. The force required to compress a spring varies directly as the change in the length of the spring. If a force of 25 pounds is required to compress a spring 4 inches, how much force is required to compress the spring 8 inches?

Let F = force and l = length
Since F varies directly as l, there is a constant k such that $F = kl$.

$$F = kl$$

$$25 = k(4)$$

$$6.25 = k$$

Now use $F = kl$ to find the value of F when $l = 8$.

$$F = 6.25(8) = 50$$

It takes 50 pounds to compress the spring 8 in.

2. The circumference of a circle varies directly as the radius. A circle with a radius of 7 centimeters has a circumference of 43.96 centimeters. Find the circumference if the radius changes to 11 centimeters.

_____

Objective 1 Practice Exercises

For extra help, see Examples 1–2 on page 565 of your text.

Solve each problem involving direct variation.

1. If y varies directly as x, and $x = 14$ when $y = 42$, find y when $x = 4$.

    1. _____

2. If c varies directly as d, and $c = 100$ when $d = 5$, find c when $d = 3$.

    2. _____

Solve the problem.

3. For a given period of time, the interest earned on an   3. _____
    investment varies directly as the interest rate. If the
    interest is $125 when the rate is 5%, find the interest
    when the rate is $6\frac{1}{2}\%$.

Objective 2   Solve inverse variation problems.

Video Examples

Review these examples for Objective 2:

3. If g varies inversely as f, and $g = 6$ when $f = 12$,
    find g when $f = 18$.

    Since g varies inversely as f, there is a

    constant k such that $g = \dfrac{k}{f}$. We know that

    $g = 6$ when $f = 12$, so we can find k.

    $$g = \frac{k}{f}$$

    $$6 = \frac{k}{12}$$

    $$72 = k$$

    Since $g = \dfrac{72}{f}$, we let $f = 18$ and solve for g.

    $$g = \frac{72}{f} = \frac{72}{18} = 4$$

    Therefore, when $f = 18$, $g = 4$.

4. For a specified distance, time varies inversely
    with speed. If Ramona walks a certain distance
    on a treadmill in 40 minutes at 4.2 miles per
    hour, how long will it take her to walk the same
    distance at 3.5 miles per hour?

    Let t = time and s = speed.
    Since t varies inversely as s, there is a constant k

    such that $t = \dfrac{k}{s}$. Recall that 40 min $= \dfrac{40}{60}$ hr .

Now Try:

3. If y varies inversely as x, and
    $y = 10$ when $x = 2$, find y
    when $x = 4$.

    _____

4. If the temperature is constant,
    the pressure of a gas in a
    container varies inversely as the
    volume of the container. If the
    pressure is 9 pounds per square
    foot in a container of 6 cubic
    feet, what is the pressure in a
    container of 7.5 cubic feet?

    _____

$$t = \frac{k}{s}$$

$$\frac{40}{60} = \frac{k}{4.2}$$

$$2.8 = k$$

Now use $t = \frac{k}{s}$ to find the value of t

when $s = 3.5$.

$$t = \frac{2.8}{3.5} = \frac{4}{5}$$

It takes $\frac{4}{5}$ hr, or 48 min to walk the same

distance.

Objective 2 Practice Exercises

For extra help, see Examples 3–4 on page 566 of your text.

Solve each problem involving indirect variation.

4. If y varies inversely as x, and $y = 20$ when $x = 4$,   4. _____
    find y when $x = 12$.

5. If n varies inversely as m, and $n = 10.5$ when   5. _____
    $m = 1.2$, find n when $m = 5.6$.

Solve the problem.

6. The length of a violin string varies inversely with the   6. _____
    frequency of its vibrations. A 10-inch violin string
    vibrates at a frequency of 512 cycles per second.
    Find the frequency of an 8-inch string.

Chapter 8 ROOTS AND RADICALS

8.1   Evaluating Roots

Learning Objectives
1   Find square roots.
2   Decide whether a given root is rational, irrational, or not a real number.
3   Find decimal approximations for irrational square roots.
4   Use the Pythagorean theorem.
5   Find cube, fourth, and other roots.

Key Terms

Use the vocabulary terms listed below to complete each statement in exercises 1−10.

> **square root   principal square root   radicand**
>
> **radical   radical expression   perfect square**
>
> **irrational number   cube root   index (order)   perfect cube**

1. The number or expression inside a radical sign is called the _____.

2. A number with a rational square root is called a _____.

3. In a radical of the form $\sqrt[n]{a}$, the number n is the _____.

4. The number b is a _____ of a if $b^2 = a$.

5. The expression $\sqrt[n]{a}$ is called a _____.

6. The positive square root of a number is its _____.

7. A number with a rational cube root is called a _____.

8. A real number that is not rational is called an _____.

9. A _____ is a radical sign and the number or expression in it.

10. The number b is a _____ of a if $b^3 = a$.

Objective 1   Find square roots.

Video Examples

Review these examples for Objective 1:

1. Find the square roots of 64.

    What number multiplied by itself equals 64?

    $8^2 = 64$ and $(-8)^2 = 64$.

    Thus, 64 has two square roots: 8 and −8.

Now Try:

1. Find the square roots of 81.

    _____

2. Find each square root.

   a. $\sqrt{121}$

   $11^2 = 121$, so $\sqrt{121} = 11$.

   d. $\sqrt{0.36}$

   $\sqrt{0.36} = 0.6$

3. Find the square of each radical expression.

   a. $\sqrt{17}$

   The square of $\sqrt{17}$ is $\left(\sqrt{17}\right)^2 = 17$.

   c. $\sqrt{w^2 + 3}$

   $\left(\sqrt{w^2 + 3}\right)^2 = w^2 + 3$

2. Find each square root.

   a. $\sqrt{169}$

   _____

   d. $\sqrt{0.64}$

   _____

3. Find the square of each radical expression.

   a. $\sqrt{19}$

   _____

   c. $\sqrt{n^2 + 5}$

   _____

Objective 1 Practice Exercises

For extra help, see Examples 1–3 on pages 584–585 of your text.

Find all square roots of each number.

   1.   625

   1. _____

   2. $\dfrac{121}{196}$

   2. _____

Find the square root.

   3. $\sqrt{\dfrac{900}{49}}$

   3. _____

Objective 2   Decide whether a given root is rational, irrational, or not a real number.

Video Examples

Review these examples for Objective 2:

4. Tell whether each square root is rational, irrational, or not a real number.

   a. $\sqrt{5}$

   Because 5 is not a perfect square, $\sqrt{5}$ is irrational.

Now Try:

4. Tell whether each square root is rational, irrational, or not a real number.

   a. $\sqrt{11}$

   _____

b. $\sqrt{81}$

81 is a perfect square, 9^2, so $\sqrt{81} = 9$ is a rational number.

c. $\sqrt{-16}$

There is no real number whose square is –16. Therefore, $\sqrt{-16}$ is not a real number.

b. $\sqrt{100}$

_____

c. $\sqrt{-9}$

_____

Objective 2 Practice Exercises

For extra help, see Example 4 on page 586 of your text.

Tell whether each square root is rational, irrational, *or* not a real number.

4. $\sqrt{72}$

5. $\sqrt{-36}$

6. $\sqrt{6400}$

4. _____

5. _____

6. _____

Objective 3   Find decimal approximations for irrational square roots.

Video Examples

Review this example for Objective 3:
5c. Find a decimal approximation for the square root. Round answers to the nearest thousandth.

$-\sqrt{596}$

$-\sqrt{596} \approx -24.413$

Now Try:
5c. Find a decimal approximation for the square root. Round answers to the nearest thousandth.

$-\sqrt{678}$

_____

Objective 3 Practice Exercises

For extra help, see Example 5 on page 587 of your text.

Use a calculator to find a decimal approximation for each square root. Round answers to the nearest thousandth.

7. $\sqrt{32}$

8. $-\sqrt{131}$

9. $\sqrt{210}$

7. _____

8. _____

9. _____

Objective 4   Use the Pythagorean theorem.

Video Examples

Review these examples for Objective 4:

6a. Find the length of the unknown side of the right triangle with sides a, b, and c, where c is the hypotenuse.

$a = 5$, $b = 12$

Use the Pythagorean theorem.

$$a^2 + b^2 = c^2$$
$$5^2 + 12^2 = c^2$$
$$25 + 144 = c^2$$
$$169 = c^2$$

Since the length of a side of a triangle must be a positive number, find the positive square root of 169 to get c.

$$c = \sqrt{169} = 13$$

7. A ladder 25 feet long leans against a wall. The foot of the ladder is 7 feet from the base of the wall. How high up the wall does the top of the ladder rest?

Step 1 Read the problem again.

Step 2 Assign a variable. Let a represent the height of the top of the ladder when measured straight down to the ground.

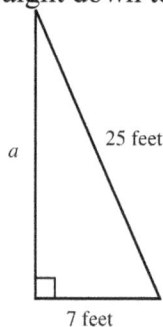

Step 3 Write an equation using the Pythagorean theorem.

$$a^2 + b^2 = c^2$$
$$a^2 + 7^2 = 25^2$$

Step 4 Solve.

Now Try:

6a. Find the length of the unknown side of the right triangle with sides a, b, and c, where c is the hypotenuse.
$a = 8$, $b = 15$

_____

7. Susan started to drive due south at the same time John started to drive due west. John drove 21 miles in the same time that Susan drove 28 miles. How far apart were they at that time?

_____

$$a^2 + 49 = 625$$

$$a^2 = 576$$

$$a = 24$$

Choose the positive square root of 576 since a represents a length.

Step 5 State the answer. The top of the ladder rests 24 ft up the wall.

Step 6 Check. From the figure, we have the following.

$$24^2 + 7^2 \overset{?}{=} 25^2$$

$$576 + 49 = 625$$

The check confirms that the top of the ladder rests 24 ft up the wall.

Objective 4 Practice Exercises

For extra help, see Examples 6–7 on pages 587–588 of your text.

Find the length of the unknown side of each right triangle with sides a, b, and c, where c is the hypotenuse. If necessary, round your answer to the nearest thousandth.

10. $a = 5,\ b = 9$ 10. _____

11. $c = 15,\ a = 12$ 11. _____

Use the Pythagorean formula to solve the problem. If necessary, round your answer to the nearest thousandth.

12.   A plane flies due east for 35 miles and then due south until it is 37 miles from its starting point. How far south did the plane fly?

     12. _____

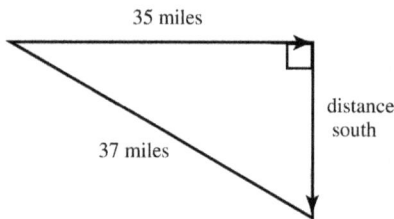

35 miles

distance
south

37 miles

Objective 5   **Find cube, fourth, and other roots.**

Video Examples

Review these examples for Objective 5:

8.   Find each cube root.

    a. $\sqrt[3]{64}$

    Because $4^3 = 64$, $\sqrt[3]{64} = 4$.

    b. $\sqrt[3]{-64}$

    $\sqrt[3]{-64} = -4$, because $(-4)^3 = -64$.

9.   Find each root.

    a. $\sqrt[4]{81}$

    $\sqrt[4]{81} = 3$, because 3 is positive and $3^4 = 81$.

    e. $\sqrt[5]{-1024}$

    $\sqrt[5]{-1024} = -4$, because $(-4)^5 = -1024$.

Now Try:

8.   Find each cube root.

    a. $\sqrt[3]{125}$

    _____

    b. $\sqrt[3]{-125}$

    _____

9.   Find each root.

    a. $\sqrt[4]{625}$

    _____

    e. $\sqrt[5]{-3125}$

    _____

Name: _____ Date: _____

Instructor: _____ Section: _____

Objective 5 Practice Exercises

For extra help, see Examples 8–9 on pages 589–590 of your text.

Find each root.

13. $\sqrt[3]{-343}$ 13. _____

14. $\sqrt[4]{256}$ 14. _____

15. $\sqrt[7]{-1}$ 15. _____

Chapter 8 ROOTS AND RADICALS

8.2   Multiplying, Dividing, and Simplifying Radicals

Learning Objectives
1   Multiply square root radicals.
2   Simplify radicals using the product rule.
3   Simplify radicals using the quotient rule.
4   Simplify radicals involving variables.
5   Simplify other roots.

Key Terms

Use the vocabulary terms listed below to complete each statement in exercises 1−3.

perfect cube   radical   radicand

1.   A root of a number is called a _____.

2.   A number with a rational cube root is called a _____.

3.   The _____ is the number or expression inside a radical sign.

Objective 1   Multiply square root radicals.

Video Examples

Review these examples for Objective 1:

1.   Use the product rule for radicals to find each product.

     b. $\sqrt{3} \cdot \sqrt{13}$

     $\sqrt{3} \cdot \sqrt{13} = \sqrt{39}$

     c. $\sqrt{17} \cdot \sqrt{b}$ $(b \geq 0)$

     $\sqrt{17} \cdot \sqrt{b} = \sqrt{17b}$

Now Try:

1.   Use the product rule for radicals to find each product.

     b. $\sqrt{6} \cdot \sqrt{7}$

     _____

     c. $\sqrt{15} \cdot \sqrt{c}$ $(c \geq 0)$

     _____

Objective 1 Practice Exercises

For extra help, see Example 1 on page 595 of your text.

Use the product rule for radicals to find each product.

1. $\sqrt{13} \cdot \sqrt{5}$ 1. _____

2. $\sqrt{11} \cdot \sqrt{2}$ 2. _____

3. $\sqrt{3x} \cdot \sqrt{7}, \ x > 0$ 3. _____

Objective 2   Simplify radicals using the product rule.

Video Examples

Review these examples for Objective 2:   **Now Try:**

2c. Simplify each radical.   2c. Simplify each radical.

   $\sqrt{48}$ $\sqrt{80}$

   $\sqrt{48} = \sqrt{16 \cdot 3} = \sqrt{16} \cdot \sqrt{3} = 4\sqrt{3}$ _____

3c. Find the product and simplify.   3. Find the product and simplify.

   $7\sqrt{10} \cdot 4\sqrt{5}$ $4\sqrt{5} \cdot 3\sqrt{10}$

   $\begin{aligned} 7\sqrt{10} \cdot 4\sqrt{5} &= 7 \cdot 4\sqrt{10 \cdot 5} \\ &= 28\sqrt{50} \\ &= 28\sqrt{25 \cdot 2} \\ &= 28\sqrt{25} \cdot \sqrt{2} \\ &= 28 \cdot 5\sqrt{2} \\ &= 140\sqrt{2} \end{aligned}$ _____

Objective 2 Practice Exercises

For extra help, see Examples 2–3 on pages 595–597 of your text.

Simplify the radical.

   4. $\sqrt{405}$ 4. _____

Find each product and simplify.

   5. $\sqrt{11} \cdot \sqrt{33}$ 5. _____

6. $\sqrt{18} \cdot \sqrt{24}$ **6.** _____

Objective 3   Simplify radicals using the quotient rule.

Video Examples

Review these examples for Objective 3:

4. Use the quotient rule to simplify each radical.

a. $\sqrt{\dfrac{121}{4}}$

$\sqrt{\dfrac{121}{4}} = \dfrac{\sqrt{121}}{\sqrt{4}} = \dfrac{11}{2}$

b. $\dfrac{\sqrt{75}}{\sqrt{3}}$

$\dfrac{\sqrt{75}}{\sqrt{3}} = \sqrt{\dfrac{75}{3}} = \sqrt{25} = 5$

c. $\sqrt{\dfrac{5}{16}}$

$\sqrt{\dfrac{5}{16}} = \dfrac{\sqrt{5}}{\sqrt{16}} = \dfrac{\sqrt{5}}{4}$

5. Simplify.

$\dfrac{32\sqrt{30}}{8\sqrt{5}}$

$\dfrac{32\sqrt{30}}{8\sqrt{5}} = \dfrac{32}{8} \cdot \dfrac{\sqrt{30}}{\sqrt{5}}$

$\qquad = \dfrac{32}{8} \cdot \sqrt{\dfrac{30}{5}}$

$\qquad = 4\sqrt{6}$

Now Try:

4. Use the quotient rule to simplify each radical.

a. $\sqrt{\dfrac{169}{9}}$

_____

b. $\dfrac{\sqrt{405}}{\sqrt{5}}$

_____

c. $\sqrt{\dfrac{7}{64}}$

_____

5. Simplify.

$\dfrac{9\sqrt{30}}{3\sqrt{3}}$

_____

6. Simplify.

$$\sqrt{\frac{5}{7}} \cdot \sqrt{\frac{2}{7}}$$

$$\sqrt{\frac{5}{7}} \cdot \sqrt{\frac{2}{7}} = \sqrt{\frac{5}{7} \cdot \frac{2}{7}}$$

$$= \sqrt{\frac{10}{49}}$$

$$= \frac{\sqrt{10}}{\sqrt{49}}, \quad \text{or} \quad \frac{\sqrt{10}}{7}$$

6. Simplify.

$$\sqrt{\frac{7}{6}} \cdot \sqrt{\frac{1}{6}}$$

_____

Objective 3 Practice Exercises

For extra help, see Examples 4–6 on pages 597–598 of your text.

Use the quotient rule and product rule, as necessary to simplify each expression.

7. $\sqrt{\dfrac{25}{81}}$

7. _____

8. $\dfrac{\sqrt{24}}{2\sqrt{6}}$

8. _____

9. $\sqrt{\dfrac{2}{125}} \cdot \sqrt{\dfrac{2}{5}}$

9. _____

Objective 4   Simplify radicals involving variables.

Video Examples

Review this example for Objective 4:

7c. Simplify the radical. Assume that all variables represent nonnegative real numbers.

$$\sqrt{27q^{12}}$$

$$\sqrt{27q^{12}} = \sqrt{9 \cdot 3 \cdot q^{12}}$$
$$= \sqrt{9} \cdot \sqrt{3} \cdot \sqrt{q^{12}}$$
$$= 3 \cdot \sqrt{3} \cdot q^{6}$$
$$= 3q^{6}\sqrt{3}$$

Now Try:

7c. Simplify the radical. Assume that all variables represent nonnegative real numbers.

$$\sqrt{72x^{4}}$$

_____

Objective 4 Practice Exercises

For extra help, see Example 7 on page 599 of your text.

Simplify each radical. Assume that all variables represent positive real numbers.

10. $\sqrt{p^{2}q^{6}}$

10. _____

11. $\sqrt{32x^{4}y^{5}}$

11. _____

12. $\sqrt{\dfrac{81}{25x^{6}}}$

12. _____

Objective 5   Simplify other roots.

Video Examples

Review these examples for Objective 5: | **Now Try:**

8c. Simplify the radical.

$$\sqrt[3]{\frac{64}{27}}$$

$$\sqrt[3]{\frac{64}{27}} = \frac{\sqrt[3]{64}}{\sqrt[3]{27}} = \frac{4}{3}$$

8c. Simplify the radical.

$$\sqrt[3]{\frac{8}{343}}$$

_____

9. Simplify each radical.

   a. $\sqrt[4]{m^8}$

$$\sqrt[4]{m^8} = m^2$$

9. Simplify each radical.

   a. $\sqrt[4]{m^{12}}$

_____

   c. $\sqrt[3]{40a^5}$

$$\sqrt[3]{40a^5} = \sqrt[3]{8a^3 \cdot 5a^2}$$
$$= \sqrt[3]{8a^3} \cdot \sqrt[3]{5a^2}$$
$$= 2a\sqrt[3]{5a^2}$$

   c. $\sqrt[4]{80a^5}$

_____

Objective 5 Practice Exercises

For extra help, see Examples 8–9 on page 600 of your text.

Simplify each expression.

13. $\sqrt[5]{-64}$ **13.** _____

14. $\sqrt[3]{\frac{1728}{1000}}$ **14.** _____

15. $\sqrt[4]{\frac{625}{256}}$ **15.** _____

Chapter 8 ROOTS AND RADICALS

8.3   Adding and Subtracting Radicals

Learning Objectives	
1	Add and subtract radicals.
2	Simplify radical sums and differences.
3	Simplify more complicated radical expressions.

Key Terms

Use the vocabulary terms listed below to complete each statement in exercises 1–3.

like radicals   **index**   **unlike radicals**

1.   In the expression, $\sqrt[4]{x^2}$, the "4" is called the _____.

2.   The expressions $2\sqrt{2}$ and $6\sqrt[3]{2}$ are _____.

3.   The expressions $2\sqrt{2}$ and $7\sqrt{2}$ are _____.

Objective 1   Add and subtract radicals.

Video Examples

Review these examples for Objective 1:

1.   Add or subtract, as indicated.

   b. $9\sqrt{13} - 12\sqrt{13}$

   $9\sqrt{13} - 12\sqrt{13} = (9-12)\sqrt{13} = -3\sqrt{13}$

   c. $\sqrt{15} + 3\sqrt{15}$

   $\sqrt{15} + 3\sqrt{15} = 1\sqrt{15} + 3\sqrt{15}$
   $\qquad\qquad = (1+3)\sqrt{15}$
   $\qquad\qquad = 4\sqrt{15}$

Now Try:

1.   Add or subtract, as indicated.

   b. $7\sqrt{17} - 13\sqrt{17}$

   _____

   c. $\sqrt{14} + 5\sqrt{14}$

   _____

Objective 1 Practice Exercises

For extra help, see Example 1 on page 605 of your text.

Add or subtract wherever possible.

1. $5\sqrt{2} + \sqrt{3}$

1. _____

2. $6\sqrt{3} - 2\sqrt{3} + 4\sqrt{3}$ 2. _____

3. $3\sqrt{5} - 9\sqrt{5} + \sqrt{5}$ 3. _____

Objective 2   Simplify radical sums and differences.

Video Examples

Review these examples for Objective 2:	Now Try:
2. Add or subtract, as indicated.	2. Add or subtract, as indicated.

b. $\sqrt{40} - \sqrt{50}$ **b.** $\sqrt{72} - \sqrt{50}$

$$\sqrt{40} - \sqrt{50} = \sqrt{4\cdot10} - \sqrt{25\cdot2}$$
$$= \sqrt{4}\cdot\sqrt{10} - \sqrt{25}\cdot\sqrt{2}$$
$$= 2\sqrt{10} - 5\sqrt{2}$$ _____

c. $2\sqrt{28} + 8\sqrt{63}$ **c.** $3\sqrt{24} + 7\sqrt{54}$

$$2\sqrt{28} + 8\sqrt{63} = 2\left(\sqrt{4}\cdot\sqrt{7}\right) + 8\left(\sqrt{9}\cdot\sqrt{7}\right)$$
$$= 2\left(2\sqrt{7}\right) + 8\left(3\sqrt{7}\right)$$
$$= 4\sqrt{7} + 24\sqrt{7}$$ _____
$$= 28\sqrt{7}$$

Objective 2 Practice Exercises

For extra help, see Example 2 on pages 605–606 of your text.

Simplify and add or subtract wherever possible.

4. $4\sqrt{128} + 2\sqrt{32}$ 4. _____

5. $7\sqrt{162} - 9\sqrt{32}$ **5.** _____

6. $5\sqrt{32} - 8\sqrt{18} + 2\sqrt{20}$ **6.** _____

Objective 3   Simplify more complicated radical expressions.

Video Examples

Review these examples for Objective 3:

3. Simplify each radical expression. Assume that all variables represent nonnegative real numbers.

   b. $\sqrt{75k} + \sqrt{108k}$

$$\sqrt{75k} + \sqrt{108k} = \sqrt{25 \cdot 3k} + \sqrt{36 \cdot 3k}$$
$$= \sqrt{25} \cdot \sqrt{3k} + \sqrt{36} \cdot \sqrt{3k}$$
$$= 5\sqrt{3k} + 6\sqrt{3k}$$
$$= 11\sqrt{3k}$$

   c. $7x\sqrt{20} + 4\sqrt{5x^2}$

$$7x\sqrt{20} + 4\sqrt{5x^2} = 7x\sqrt{4 \cdot 5} + 4\sqrt{5 \cdot x^2}$$
$$= 7x\sqrt{4} \cdot \sqrt{5} + 4\sqrt{5} \cdot \sqrt{x^2}$$
$$= 7x \cdot 2\sqrt{5} + 4x\sqrt{5}$$
$$= 14x\sqrt{5} + 4x\sqrt{5}$$
$$= 18x\sqrt{5}$$

Now Try:

3. Simplify each radical expression. Assume that all variables represent nonnegative real numbers.

   b. $\sqrt{28k} + \sqrt{63k}$

_____

   c. $6x\sqrt{75} + 2\sqrt{3x^2}$

_____

d. $8\sqrt[3]{81x^3} - \sqrt[3]{375x^3}$

$8\sqrt[3]{81x^3} - \sqrt[3]{375x^3}$

$= 8\sqrt[3]{(27x^3)(3)} - \sqrt[3]{(125x^3)(3)}$

$= 8(3x)\sqrt[3]{3} - 5x\sqrt[3]{3}$

$= 24x\sqrt[3]{3} - 5x\sqrt[3]{3}$

$= 19x\sqrt[3]{3}$

d. $7\sqrt[3]{16m^3} - \sqrt[3]{54m^3}$

_____

Objective 3 Practice Exercises

For extra help, see Example 3 on pages 606–607 of your text.

Perform the indicated operations. Assume that all variables represent nonnegative real numbers.

7. $\sqrt{5} \cdot \sqrt{7} + 3\sqrt{35}$

7. _____

8. $3\sqrt{125x} - \sqrt{80x} + 2\sqrt{45x}$

8. _____

9. $11\sqrt{5w} \cdot \sqrt{30w} - 8w\sqrt{24}$

9. _____

Chapter 8 ROOTS AND RADICALS

8.4   Rationalizing the Denominator

Learning Objectives	
1	Rationalize denominators with square roots.
2	Write radicals in simplified form.
3	Rationalize denominators with cube roots.

Key Terms

Use the vocabulary terms listed below to complete each statement in exercises 1−3.

rationalizing the denominator   **product rule**   **quotient rule**

1. The _____ states that $\sqrt{a} \cdot \sqrt{b} = \sqrt{ab}$.

2. The process of _____ is changing the denominator of a fraction from a radical to an expression not involving a radical.

3. The _____ states that $\sqrt{\dfrac{a}{b}} = \dfrac{\sqrt{a}}{\sqrt{b}}$.

Objective 1   Rationalize denominators with square roots.

Video Examples

Review these examples for Objective 1:

1. Rationalize each denominator.

   a. $\dfrac{10}{\sqrt{5}}$

   $\dfrac{10}{\sqrt{5}} = \dfrac{10 \cdot \sqrt{5}}{\sqrt{5} \cdot \sqrt{5}}$ Multiply by $\dfrac{\sqrt{5}}{\sqrt{5}} = 1$.

   $\quad = \dfrac{10\sqrt{5}}{5}$

   $\quad = 2\sqrt{5}$ Write in lowest terms.

   b. $\dfrac{18}{\sqrt{27}}$

   $\dfrac{18}{\sqrt{27}} = \dfrac{18}{3\sqrt{3}}$

   $\quad = \dfrac{18 \cdot \sqrt{3}}{3\sqrt{3} \cdot \sqrt{3}}$ Multiply by $\dfrac{\sqrt{3}}{\sqrt{3}} = 1$.

   $\quad = \dfrac{18\sqrt{3}}{3 \cdot 3}$

   $\quad = 2\sqrt{3}$

Now Try:

1. Rationalize each denominator.

   a. $\dfrac{14}{\sqrt{7}}$

   _____

   b. $\dfrac{5}{\sqrt{75}}$

   _____

Objective 1 Practice Exercises

For extra help, see Example 1 on page 610 of your text.

Rationalize each denominator.

1. $\dfrac{15}{\sqrt{10}}$

    1. _____

2. $\dfrac{6}{\sqrt{28}}$

    2. _____

3. $\dfrac{3\sqrt{5}}{\sqrt{125}}$

    3. _____

Objective 2   Write radicals in simplified form.

Video Examples

Review these examples for Objective 2:

2. Simplify.

$$\sqrt{\dfrac{75}{7}}$$

$$\sqrt{\dfrac{75}{7}} = \dfrac{\sqrt{75} \cdot \sqrt{7}}{\sqrt{7} \cdot \sqrt{7}}$$

$$= \dfrac{\sqrt{25 \cdot 3} \cdot \sqrt{7}}{7}$$

$$= \dfrac{\sqrt{25} \cdot \sqrt{3} \cdot \sqrt{7}}{7}$$

$$= \dfrac{5\sqrt{3} \cdot \sqrt{7}}{7}$$

$$= \dfrac{5\sqrt{21}}{7}$$

Now Try:

2. Simplify.

$$\sqrt{\dfrac{7}{12}}$$

_____

3. Simplify.

$$\sqrt{\frac{7}{15}} \cdot \sqrt{\frac{1}{10}}$$

$$\sqrt{\frac{7}{15}} \cdot \sqrt{\frac{1}{10}} = \sqrt{\frac{7}{15} \cdot \frac{1}{10}}$$

$$= \sqrt{\frac{7}{150}}$$

$$= \frac{\sqrt{7}}{\sqrt{150}}$$

$$= \frac{\sqrt{7}}{\sqrt{25} \cdot \sqrt{6}}$$

$$= \frac{\sqrt{7}}{5\sqrt{6}}$$

$$= \frac{\sqrt{7} \cdot \sqrt{6}}{5\sqrt{6} \cdot \sqrt{6}}$$

$$= \frac{\sqrt{42}}{5 \cdot 6}$$

$$= \frac{\sqrt{42}}{30}$$

3. Simplify.

$$\sqrt{\frac{1}{3}} \cdot \sqrt{\frac{2}{15}}$$

_____

Objective 2 Practice Exercises

For extra help, see Examples 2–4 on pages 611–612 of your text.

Perform the indicated operations and write all answers in simplest form. Rationalize all denominators. Assume that all variables represent positive real numbers.

4. $\sqrt{\frac{3}{2}} \cdot \sqrt{\frac{5}{6}}$

4. _____

5. $\dfrac{\sqrt{k^2 m^4}}{\sqrt{k^5}}$

5. _____

6. $\sqrt{\dfrac{5a^2b^3}{6}}$ **6.** _____

Objective 3   Rationalize denominators with cube roots.

Video Examples

Review this example for Objective 3: | **Now Try:**

5b. Rationalize the denominator. | **5b.** Rationalize the denominator.

$$\dfrac{\sqrt[3]{7}}{\sqrt[3]{25}}$$ $$\dfrac{\sqrt[3]{11}}{\sqrt[3]{9}}$$

$$\dfrac{\sqrt[3]{7}}{\sqrt[3]{25}} = \dfrac{\sqrt[3]{7}\cdot\sqrt[3]{5}}{\sqrt[3]{25}\cdot\sqrt[3]{5}} = \dfrac{\sqrt[3]{35}}{\sqrt[3]{125}} = \dfrac{\sqrt[3]{35}}{5}$$

_____

Objective 3 Practice Exercises

For extra help, see Example 5 on page 613 of your text.

Rationalize each denominator. Assume that all variables in the denominator represent nonzero real numbers.

7. $\dfrac{\sqrt[3]{6}}{\sqrt[3]{9}}$ **7.** _____

8. $\sqrt[3]{\dfrac{8t}{125u}}$ **8.** _____

9. $\sqrt[3]{\dfrac{5}{49x}}$ **9.** _____

Chapter 8 ROOTS AND RADICALS

8.5   More Simplifying and Operations with Radicals

Learning Objectives
1   Simplify products of radical expressions.
2   Use conjugates to rationalize denominators of radical expressions.
3   Write radical expressions with quotients in lowest terms.

Key Terms

Use the vocabulary terms listed below to complete each statement in exercises 1−2.

> **conjugate**   **rationalize the denominator**

1.   To _____ of $\dfrac{3}{\sqrt{5}}$, multiply both the numerator and

    the denominator by $\sqrt{5}$.

2.   The _____ of $a + b$ is $a - b$.

Objective 1   Simplify products of radical expressions.

Video Examples

Review these examples for Objective 1:

1.   Find each product, and simplify.

    a. $\sqrt{7}\left(\sqrt{27} - \sqrt{48}\right)$

$$\sqrt{7}\left(\sqrt{27} - \sqrt{48}\right) = \sqrt{7}\left(3\sqrt{3} - 4\sqrt{3}\right)$$
$$= \sqrt{7}\left(-\sqrt{3}\right)$$
$$= -\sqrt{7 \cdot 3}$$
$$= -\sqrt{21}$$

    b. $\left(\sqrt{7} + 5\sqrt{2}\right)\left(\sqrt{7} - 8\sqrt{2}\right)$

Use the FOIL method to multiply.
$$\left(\sqrt{7} + 5\sqrt{2}\right)\left(\sqrt{7} - 8\sqrt{2}\right)$$
$$= \sqrt{7}\left(\sqrt{7}\right) + \sqrt{7}\left(-8\sqrt{2}\right) + 5\sqrt{2}\left(\sqrt{7}\right) + 5\sqrt{2}\left(-8\sqrt{2}\right)$$
$$= 7 - 8\sqrt{14} + 5\sqrt{14} - 40 \cdot 2$$
$$= 7 - 3\sqrt{14} - 80$$
$$= -73 - 3\sqrt{14}$$

Now Try:

1.   Find each product, and simplify.

    a. $\sqrt{2}\left(\sqrt{45} - \sqrt{20}\right)$

    _____

    b. $\left(\sqrt{6} + 3\sqrt{5}\right)\left(\sqrt{6} - 5\sqrt{5}\right)$

    _____

Name:
Instructor:

Date:
Section:

3a. Find the product.

$$\left(7+\sqrt{5}\right)\left(7-\sqrt{5}\right)$$

Recall $(a+b)(a-b)=a^2-b^2$.

$$\left(7+\sqrt{5}\right)\left(7-\sqrt{5}\right)=7^2-\left(\sqrt{5}\right)^2$$
$$=49-5$$
$$=44$$

2b. Find the product.

$$\left(3\sqrt{5}+7\right)^2$$

$$\left(3\sqrt{5}+7\right)^2=\left(3\sqrt{5}\right)^2+2\left(3\sqrt{5}\right)(7)+7^2$$
$$=45+42\sqrt{5}+49$$
$$=94+42\sqrt{5}$$

3a. Find the product.

$$\left(11+\sqrt{4}\right)\left(11-\sqrt{4}\right)$$

_____

2b. Find the product.

$$\left(5\sqrt{2}+8\right)^2$$

_____

Objective 1 Practice Exercises

For extra help, see Examples 1–3 on pages 616–618 of your text.

Find each product, and simplify.

1. $\sqrt{7}\left(2\sqrt{8}-9\sqrt{7}\right)$

1. _____

2. $\left(4\sqrt{5}+\sqrt{3}\right)\left(\sqrt{2}-\sqrt{7}\right)$

2. _____

3. $\left(2\sqrt{3}-5\sqrt{2}\right)\left(2\sqrt{3}+5\sqrt{2}\right)$

3. _____

Name: Date:

Instructor: Section:

Objective 2   Use conjugates to rationalize denominators of radical expressions.

Video Examples

Review these examples for Objective 2:

4. Simplify by rationalizing each denominator.

a. $\dfrac{7}{4+\sqrt{7}}$

$$\frac{7}{4+\sqrt{7}} = \frac{7(4-\sqrt{7})}{(4+\sqrt{7})(4-\sqrt{7})}$$

$$= \frac{7(4-\sqrt{7})}{4^2-(\sqrt{7})^2}$$

$$= \frac{7(4-\sqrt{7})}{16-7}, \quad \text{or} \quad \frac{7(4-\sqrt{7})}{9}$$

b. $\dfrac{8+\sqrt{3}}{\sqrt{3}-4}$

$$\frac{8+\sqrt{3}}{\sqrt{3}-4} = \frac{(8+\sqrt{3})(\sqrt{3}+4)}{(\sqrt{3}-4)(\sqrt{3}+4)}$$

$$= \frac{8\sqrt{3}+32+3+4\sqrt{3}}{3-16}$$

$$= \frac{12\sqrt{3}+35}{-13}$$

$$= \frac{-12\sqrt{3}-35}{13}$$

Now Try:

4. Simplify by rationalizing each denominator.

a. $\dfrac{10}{7+\sqrt{10}}$

_____

b. $\dfrac{5+\sqrt{6}}{\sqrt{6}-3}$

_____

Objective 2 Practice Exercises

For extra help, see Example 4 on pages 618–619 of your text.

Rationalize each denominator. Write quotients in lowest terms.

4. $\dfrac{\sqrt{2}}{\sqrt{5}-2}$

4. _____

5. $\dfrac{\sqrt{6}+2}{\sqrt{2}-4}$ **5.** _____

6. $\dfrac{\sqrt{5}-2}{\sqrt{3}+2}$ **6.** _____

Objective 3   Write radical expressions with quotients in lowest terms.

Video Examples

Review this example for Objective 3:   **Now Try:**

5.   Write $\dfrac{5\sqrt{2}+10}{35}$ in lowest terms.   **5.**   Write $\dfrac{4\sqrt{5}+20}{36}$ in lowest terms.

$$\dfrac{5\sqrt{2}+10}{35}=\dfrac{5\left(\sqrt{2}+2\right)}{5(7)}$$

$$=1\cdot\dfrac{\sqrt{2}+2}{7}\quad \begin{array}{l}\text{Divide out the common}\\ \text{factor; } \frac{5}{5}=1\end{array}$$

$$=\dfrac{\sqrt{2}+2}{7}$$

                                                       _____

Objective 3 Practice Exercises

For extra help, see Example 5 on page 619 of your text.

Write each quotient in lowest terms.

7. $\dfrac{3+\sqrt{27}}{9}$ **7.** _____

8. $\dfrac{12+6\sqrt{6}}{8}$

8. _____

9. $\dfrac{135\sqrt{3}+25}{5}$

9. _____

Chapter 8 ROOTS AND RADICALS

8.6   Solving Equations with Radicals

Learning Objectives
1   Solve radical equations having square root radicals.
2   Identify equations with no solutions.
3   Solve equations by squaring a binomial.
4   Solve problems using formulas that involve radicals.

Key Terms

Use the vocabulary terms listed below to complete each statement in exercises 1–2.

> **radical equation**   **extraneous solution**

1.   An _____ is a proposed solution to an equation that does not satisfy the equation.

2.   An equation with a variable in the radicand is a _____.

Objective 1   Solve radical equations having square root radicals.

Video Examples

Review these examples for Objective 1:

1.   Solve $\sqrt{p+2} = 5$.

$$\left(\sqrt{p+2}\right)^2 = 5^2$$
$$p+2 = 25$$
$$p = 23$$

Check $\sqrt{p+2} = 5$
$$\sqrt{23+2} \stackrel{?}{=} 5$$
$$\sqrt{25} \stackrel{?}{=} 5$$
$$5 = 5 \quad \text{True}$$

The solution set is $\{23\}$.

2.   Solve $4\sqrt{x} = \sqrt{x+30}$.

$$\left(4\sqrt{x}\right)^2 = \left(\sqrt{x+30}\right)^2$$
$$4^2\left(\sqrt{x}\right)^2 = \left(\sqrt{x+30}\right)^2$$
$$16x = x+30$$
$$15x = 30$$
$$x = 2$$

Now Try:

1.   Solve $\sqrt{p+5} = 4$.

_____

2.   Solve $5\sqrt{x} = \sqrt{x+72}$.

_____

Check $4\sqrt{x} = \sqrt{x+30}$

$\quad\quad\quad 4\sqrt{2} \stackrel{?}{=} \sqrt{2+30}$

$\quad\quad\quad 4\sqrt{2} \stackrel{?}{=} \sqrt{32}$

$\quad\quad\quad 4\sqrt{2} = 4\sqrt{2}$ True

The solution set is $\{2\}$.

Objective 1 Practice Exercises

For extra help, see Examples 1–2 on pages 625–626 of your text.

Solve each equation.

1. $\sqrt{3x+1} = 3$ 1. _____

2. $\sqrt{2+4k} = 3\sqrt{k}$ 2. _____

3. $\sqrt{4x+3} = \sqrt{3x+5}$ 3. _____

Name: _____   Date: _____
Instructor: _____   Section: _____

Objective 2   Identify equations with no solutions.

Video Examples

Review this example for Objective 2:

3. Solve $\sqrt{x} = -25$.

$$\left(\sqrt{x}\right)^2 = (-25)^2$$
$$x = 625$$

Check $\sqrt{x} = -25$

$\sqrt{625} \stackrel{?}{=} -25$

$25 = -25$ False

Because the statement $25 = -25$ is false, the number 625 is not a solution. It is an extraneous solution and must be rejected. There is no solution. The solution set is $\varnothing$.

Now Try:

3. Solve $\sqrt{x} = -10$.

_____

Objective 2 Practice Exercises

For extra help, see Examples 3–4 on pages 626–627 of your text.

Solve each equation.

4. $\sqrt{x+2} + 7 = 0$

4. _____

5. $\sqrt{2m+3} = 3\sqrt{m+5}$

5. _____

6. $r = \sqrt{r^2 - 6r + 12}$ **6.** _____

Objective 3   Solve equations by squaring a binomial.

Video Examples

Review these examples for Objective 3:

6. Solve $\sqrt{9x} + 2 = 4x + 1$.

$$\sqrt{9x} + 2 = 4x + 1$$
$$\sqrt{9x} = 4x - 1$$
$$\left(\sqrt{9x}\right)^2 = (4x - 1)^2$$
$$9x = 16x^2 - 8x + 1$$
$$0 = 16x^2 - 17x + 1$$
$$0 = (16x - 1)(x - 1)$$
$$16x - 1 = 0 \quad \text{or} \quad x - 1 = 0$$
$$x = \frac{1}{16} \quad \text{or} \quad x = 1$$

Check   Let $x = \frac{1}{16}$ Let $x = 1$.

$$\sqrt{9x} + 2 = 4x + 1 \qquad \sqrt{9x} + 2 = 4x + 1$$
$$\sqrt{9\left(\frac{1}{16}\right)} + 2 \overset{?}{=} 4\left(\frac{1}{16}\right) + 1 \qquad \sqrt{9(1)} + 2 \overset{?}{=} 4(1) + 1$$
$$\frac{3}{4} + 2 \overset{?}{=} \frac{1}{4} + 1 \qquad\qquad 3 + 2 \overset{?}{=} 4 + 1$$
$$\qquad\qquad\qquad\qquad\qquad 5 = 5 \quad \text{True}$$
$$\frac{11}{4} = \frac{5}{4} \quad \text{False}$$

Only 1 is a valid solution. ($\frac{1}{16}$ is extraneous.)

The solution set is $\{1\}$.

Now Try:

6. Solve $\sqrt{x + 2} + 4 = x + 6$.

_____

5. Solve $\sqrt{2x-4}=2-x$.

$$\left(\sqrt{2x-4}\right)^2=(2-x)^2$$

$$2x-4=4-4x+x^2$$

$$0=x^2-6x+8$$

$$0=(x-2)(x-4)$$

$$x-2=0 \quad \text{or} \quad x-4=0$$

$$x=2 \quad \text{or} \quad x=4$$

Check   Let $x=2$.   Let $x=4$.

$$\sqrt{2x-4}=2-x \qquad \sqrt{2x-4}=2-x$$

$$\sqrt{2(2)-4}\overset{?}{=}2-2 \qquad \sqrt{2(4)-4}\overset{?}{=}2-4$$

$$\sqrt{4-4}\overset{?}{=}0 \qquad \sqrt{8-4}\overset{?}{=}-2$$

$$0=0 \quad \text{True} \qquad \sqrt{4}\overset{?}{=}-2$$

$$2=-2 \quad \text{False}$$

Only 2 is a valid solution. (4 is extraneous.)
The solution set is $\{2\}$.

5. Solve $\sqrt{3x-5}=x-3$.

_____

7. Solve $\sqrt{40+x}=4+\sqrt{x}$.

$$\left(\sqrt{40+x}\right)^2=\left(4+\sqrt{x}\right)^2$$

$$40+x=16+8\sqrt{x}+x$$

$$24=8\sqrt{x}$$

$$3=\sqrt{x}$$

$$9=x$$

Check $\sqrt{40+x}=4+\sqrt{x}$

$$\sqrt{40+9}\overset{?}{=}4+\sqrt{9}$$

$$\sqrt{49}\overset{?}{=}4+3$$

$$7=7 \quad \text{True}$$

The solution set is $\{9\}$.

7. Solve $\sqrt{p+4}-\sqrt{p-1}=1$.

_____

Objective 3 Practice Exercises

For extra help, see Examples 5–7 on pages 628–629 of your text.

Solve each equation.

7. $\sqrt{b-4}=b-6$ 7. _____

8. $3\sqrt{p+6}=p+6$ 8. _____

9. $q-1=\sqrt{q^2-4q+7}$ 9. _____

Objective 4   Solve problems using formulas that involve radicals.

Video Examples

Review this example for Objective 4:
8.   The sides of a triangle have lengths 84 ft, 68 ft, and 40 ft. Find the area of the triangle.

   First find the semiperimeter, *s*.
   $$s=\frac{1}{2}(a+b+c)$$
   $$s=\frac{1}{2}(84+68+40)=96$$
   Now use Heron's formula to find the area.

Now Try:
8.   The sides of a triangle have lengths 90 m, 153 m, and 189 m. Find the area of the triangle.

   _____

$$A = \sqrt{s(s-a)(s-b)(s-c)}$$

$$A = \sqrt{96(96-84)(96-68)(96-40)}$$

$$A = \sqrt{96(12)(28)(56)}$$

$$A = \sqrt{1,806,336}$$

$$A = 1344 \text{ ft}^2$$

The area is 1344 ft^2.

Objective 4 Practice Exercises

For extra help, see Example 8 on page 630 of your text.

Use Heron's formula $\mathcal{A} = \sqrt{s(s-a)(s-b)(s-c)}$ *to find the area of the triangle with the given sides. Round answers to the nearest whole number.*

10. $a = 10$ in., $b = 10$ in., $c = 12$ in.   **10.** _____

11. $a = 42.3$ cm, $b = 29.8$ cm, $c = 33.7$ cm   **11.** _____

12. $a = 23$ m, $b = 19$ m, $c = 12$ m   **12.** _____

Chapter 9 QUADRATIC EQUATIONS

9.1   Solving Quadratic Equations by the Square Root Property

Learning Objectives
1   Review the zero-factor property.
2   Solve equations of the form $x^2 = k$, where $k > 0$.
3   Solve equations of the form $(ax + b)^2 = k$, where $k > 0$.
4   Use formulas involving second-degree variables.

Key Terms

Use the vocabulary terms listed below to complete each statement in exercises 1−2.

quadratic equation   zero-factor property

1.   An equation that can be written in the form $ax^2 + bx + c = 0$ is a

_____.

2.   The _____ states that if a product equals 0, then at least one of the factors of the product also equals zero.

Objective 1   Review the zero-factor property.

Video Examples

Review this example for Objective 1:

1a. Solve each equation by the zero-factor property.

$$x^2 + 5x + 4 = 0$$

$$x^2 + 5x + 4 = 0$$
$$(x + 4)(x + 1) = 0$$
$$x + 4 = 0 \quad \text{or} \quad x + 1 = 0$$
$$x = -4 \quad \text{or} \qquad x = -1$$

The solution set is $\{-4, -1\}$.

Now Try:

1a. Solve each equation by the zero-factor property.

$$x^2 + 8x + 7 = 0$$

_____

Objective 1 Practice Exercises

For extra help, see Example 1 on page 648 of your text.

Solve each equation by using the zero-factor property.

1. $x^2 + 6x + 8 = 0$

1. _____

2. $x^2 - 196 = 0$

2. _____

3. $x^2 + 2x - 35 = 0$

3. _____

Objective 2   Solve equations of the form $x^2 = k$, where $k > 0$.

Video Examples

Review these examples for Objective 2:	Now Try:
2. Solve each equation. Write radicals in simplified form.	**2.** Solve each equation. Write radicals in simplified form.

a. $x^2 = 36$

By the square root property, if $x^2 = 36$, then
$x = \sqrt{36} = 6$ or $x = -\sqrt{36} = -6$
The solution set is {−6, 6}, or {±6}.

a. $x^2 = 81$

_____

b. $z^2 = 13$

The solutions are $z = \sqrt{13}$ or $z = -\sqrt{13}$
The solution set is $\left\{-\sqrt{13}, \sqrt{13}\right\}$, or $\left\{\pm\sqrt{13}\right\}$.

b. $z^2 = 23$

_____

3. Solve $p^2 = -64$

Because −64 is a negative number and because
the square of a real number cannot be negative,
there is no real number solution of this equation.
The solution set is ∅.

3. Solve $n^2 = -25$

_____

2c. Solve the equation. Write radicals in simplified form.

$$7m^2 - 17 = 67$$

$$7m^2 - 17 = 67$$

$$7m^2 = 84$$

$$m^2 = 12$$

$$m = \sqrt{12} \quad \text{or} \quad m = -\sqrt{12}$$

$$m = 2\sqrt{3} \quad \text{or} \quad m = -2\sqrt{3}$$

The solution set is $\left\{-2\sqrt{3},\ 2\sqrt{3}\right\}$, or $\left\{\pm 2\sqrt{3}\right\}$.

2c. Solve the equation. Write radicals in simplified form.

$$3m^2 - 15 = 210$$

_____

Objective 2 Practice Exercises

For extra help, see Examples 2–3 on pages 649–650 of your text.

Solve each equation by using the square root property. Express all radicals in simplest form.

4. $r^2 = 900$

4. _____

5. $s^2 - 98 = 0$

5 _____

6. $p^2 = -144$

6. _____

Objective 3   Solve equations of the form $(ax+b)^2 = k$, **where** $k > 0$.

Video Examples

Review these examples for Objective 3:

4b. Solve the equation.

$$(x+2)^2 = 7$$

$$x+2 = \sqrt{7} \quad \text{or} \quad x+2 = -\sqrt{7}$$

$$x = -2+\sqrt{7} \quad \text{or} \quad x = -2-\sqrt{7}$$

Check

$$\left(-2+\sqrt{7}+2\right)^2 = \left(\sqrt{7}\right)^2 = 7$$

$$\left(-2-\sqrt{7}+2\right)^2 = \left(-\sqrt{7}\right)^2 = 7$$

The solution set is $\{-2+\sqrt{7}, -2-\sqrt{7}\}$, or $\{-2\pm\sqrt{7}\}$.

5. Solve $(5r-3)^2 = 12$.

$$5r-3 = \sqrt{12} \quad \text{or} \quad 5r-3 = -\sqrt{12}$$

$$5r-3 = 2\sqrt{3} \quad \text{or} \quad 5r-3 = -2\sqrt{3}$$

$$5r = 3+2\sqrt{3} \quad \text{or} \quad 5r = 3-2\sqrt{3}$$

$$r = \frac{3+2\sqrt{3}}{5} \quad \text{or} \quad r = \frac{3-2\sqrt{3}}{5}$$

Check

$$(5r-3)^2 = 12$$

$$\left[5 \cdot \frac{3+2\sqrt{3}}{5} - 3\right]^2 \overset{?}{=} 12$$

$$\left(3+2\sqrt{3}-3\right)^2 \overset{?}{=} 12$$

$$\left(2\sqrt{3}\right)^2 \overset{?}{=} 12$$

$$12 = 12 \quad \text{True}$$

The check of the other solution is similar. The solution set is

The solution set is $\left\{\dfrac{3+2\sqrt{3}}{5}, \dfrac{3-2\sqrt{3}}{5}\right\}$.

6. Solve $(x+2)^2 = -16$.

Because the square root of -16 is not a real number, there is no real number solution for this equation. The solution set is $\varnothing$.

Now Try:

4b. Solve the equation.

$$(x+3)^2 = 13$$

_____

5. Solve $(7r-3)^2 = 32$.

_____

6. Solve $(x-7)^2 = -49$.

_____

Name: _____ Date: _____

Instructor: _____ Section: _____

Objective 3 Practice Exercises

For extra help, see Examples 4–6 on pages 650–651 of your text.

Solve each equation by using the square root property. Express all radicals in simplest form.

7. $(y+2)^2 = 16$

7. _____

8. $(7p-4)^2 = 289$

8. _____

9. $(10m-5)^2 - 9 = 0$

9. _____

Objective 4   Use formulas involving second-degree variables.

Video Examples

Review this example for Objective 4:

7. We can approximate the weight of a bass, in pounds, given its length L and its girth (distance around) g, where both are measured in inches, using the following formula.

$$w = \frac{L^2 g}{1200}$$

Approximate the length of a bass weighing 2.40 lb and having girth 9 in.

Start with the formula and substitute 2.40 for w and 9 for g.

$$w = \frac{L^2 g}{1200}$$

$$2.40 = \frac{L^2 \cdot 9}{1200}$$

$$2880 = 9L^2$$

$$L^2 = 320$$

$$L = \sqrt{320} \quad \text{or} \quad L = -\sqrt{320}$$

The calculator shows that $\sqrt{320} \approx 17.89$, so the length of the bass is almost 18 in. (We discard the solution $-\sqrt{320} \approx -17.89$, since L represents length.)

Now Try:

7. We can approximate the weight of a bass, in pounds, given its length L and its girth (distance around) g, where both are measured in inches, using the following formula.

$$w = \frac{L^2 g}{1200}$$

Approximate the length of a bass weighing 2.50 lb and having girth 10.5 in.

_____

Objective 4 Practice Exercises

For extra help, see Example 7 on page 651 of your text.

Solve each problem.

10.   The formula $A = P(1+r)^2$ gives the amount A that P dollars invested at an annual rate of interest r will grow to in two years. Mary invests $1500, and after two years, she has $1653.75. What is the interest rate?

10.   _____

11. The volume of a cylinder is given by the formula
$V = \pi r^2 h,$ where $V =$ the volume, $r =$ the radius of
the base of the cylinder, and $h =$ the height of the
cylinder. If the volume of a can is 20π in.^{3}, and its
height is 5 inches, find the radius of the can.

11. _____

12. One leg of a right triangle has length 5 cm, and the
hypotenuse has length 10 cm. Find the length of the
other leg. (Hint: Use the Pythagorean theorem.)

12. _____

Chapter 9 QUADRATIC EQUATIONS

9.2   Solving Quadratic Equations by Completing the Square

Learning Objectives
1   Solve quadratic equations by completing the square when the coefficient of the second-degree term is 1.
2   Solve quadratic equations by completing the square when the coefficient of the second-degree term is not 1.
3   Simplify the terms of an equation before solving.
4   Solve applied problems that require quadratic equations.

Key Terms

Use the vocabulary terms listed below to complete each statement in exercises 1−3.

completing the square   perfect square trinomial   square root property

1.   A _____ can be written in the form $x^2 + 2kx + k^2$ or

   $x^2 - 2kx + k^2$

2.   The _____ says that, if k is positive and $a^2 = k$, then

   $a = \pm\sqrt{k}$.

3.   Use the process called _____ in order to rewrite an equation
   so it can be solved using the square root property.

**Objective 1   Solve quadratic equations by completing the square when the coefficient
                 of the second-degree term is 1.**

Video Examples

Review these examples for Objective 1:

1.   Complete each trinomial so that it is a perfect
     square. Then factor the trinomial.

   a. $x^2 + 24x +$ _____

   The perfect square trinomial will have the form
   $x^2 + 2kx + k^2$. Thus, the middle term 24x, must
   equal 2kx.

   $24x = 2kx$

   $12 = k$

   Therefore, $k = 12$ and $k^2 = 12^2 = 144$. The

   perfect square trinomial is $x^2 + 24x + 144$,

   which factors as $(x + 12)^2$.

Now Try:

1.   Complete each trinomial so that
     it is a perfect square. Then
     factor the trinomial.

   a. $x^2 + 10x +$ _____

   _____

b. $x^2 - 36x + $ _____

The perfect square trinomial will have the form $x^2 - 2kx + k^2$. Thus, the middle term $-36x$, must equal $-2kx$.
$$-36x = -2kx$$
$$18 = k$$

Therefore, $k = 18$ and $k^2 = 18^2 = 324$. The required perfect square trinomial is $x^2 - 36x + 324$, which factors as $(x - 18)^2$.

2. Solve $x^2 + 8x + 3 = 0$.

$$x^2 + 8x + 3 = 0$$
$$x^2 + 8x = -3$$

The expression on the left must be written as a perfect square trinomial in the form $x^2 + 2kx + k^2$.
$$x^2 + 8x + \underline{}$$

Here, $2kx = 8x$, so $k = 4$ and $k^2 = 16$. The required perfect square trinomial is $x^2 + 8x + 16$ which factors as $(x + 4)^2$.

Therefore, if we add 16 to each side of $x^2 + 8x = -3$, the equation will have a perfect square trinomial on the left side, as needed.
$$x^2 + 8x + 16 = -3 + 16$$
$$(x + 4)^2 = 13$$

Use the square root property.
$$x + 4 = \sqrt{13} \qquad \text{or} \quad x + 4 = -\sqrt{13}$$
$$x = -4 + \sqrt{13} \quad \text{or} \qquad x = -4 - \sqrt{13}$$

Check by substituting $-4 + \sqrt{13}$ and then $-4 - \sqrt{13}$ for x in the original equation. The solution set is $\{-4 + \sqrt{13}, \ -4 - \sqrt{13}\}$.

3. Complete the square to solve $x^2 - 18x = 11$.

To complete the square on $x^2 - 18x$, take half the coefficient of x and square it.
$$\frac{1}{2}(-18) = -9 \quad \text{and} \quad (-9)^2 = 81$$

b. $x^2 - 22x + $ _____

_____

2. Solve $x^2 + 10x - 7 = 0$.

_____

3. Complete the square to solve $x^2 + 16x = 9$.

_____

Add the result, 81, to each side of the equation.

$$x^2 - 18x = 11$$

$$x^2 - 18x + 81 = 11 + 81$$

$$(x - 9)^2 = 92$$

$$x - 9 = \sqrt{92} \qquad \text{or} \quad x - 9 = -\sqrt{92}$$

$$x - 9 = 2\sqrt{23} \qquad\qquad x - 9 = -2\sqrt{23}$$

$$x = 9 + 2\sqrt{23} \quad \text{or} \qquad x = 9 - 2\sqrt{23}$$

A check indicates the solution set is

$$\left\{9 + 2\sqrt{23},\ 9 - 2\sqrt{23}\right\}$$

Objective 1 Practice Exercises

For extra help, see Examples 1–3 on pages 655–656 of your text.

Solve each equation by completing the square.

1. $r^2 + 8r = -4$

1. _____

2. $x^2 - 4x = 2$

2. _____

3. $x^2 + 2x = 63$

3. _____

Objective 2   Solve quadratic equations by completing the square when the coefficient of the second-degree term is not 1.

Video Examples

Review these examples for Objective 2:	**Now Try:**
4. Solve $9x^2 + 18x = 7$.	**4.** Solve $16x^2 - 64x = -55$.

Step 1 Before completing the square, the coefficient of x^2 must be 1, not 9. Divide each side by 9.

$$x^2 + 2x = \frac{7}{9}$$

Step 2 The equation is already in the correct form, with the variable terms on one side and the constant on the other.

Step 3 Complete the square. Take half the coefficient of x, and square it.

$$\frac{1}{2}(2) = 1 \quad \text{and} \quad 1^2 = 1$$

$$x^2 + 2x + 1 = \frac{7}{9} + 1$$

$$(x+1)^2 = \frac{16}{9}$$

Step 4 Solve the equation by using the square root property.

$$x + 1 = \sqrt{\frac{16}{9}} \quad \text{or} \quad x + 1 = -\sqrt{\frac{16}{9}}$$

$$x + 1 = \frac{4}{3} \qquad\qquad x + 1 = -\frac{4}{3}$$

$$x = \frac{1}{3} \quad \text{or} \quad x = -\frac{7}{3}$$

The two solutions $-\frac{7}{3}$ and $\frac{1}{3}$ check, so the solution set is $\left\{-\frac{7}{3}, \frac{1}{3}\right\}$.

6. Solve $5p^2 - 20p + 21 = 0$.

$$5p^2 - 20p + 21 = 0$$

$$p^2 - 4p + \frac{21}{5} = 0$$

$$p^2 - 4p = -\frac{21}{5}$$

The coefficient of p is -4. Take half of -4,

6. Solve $6p^2 - 36p + 55 = 0$.

square the result, and add it to each side.

$$p^2 - 4p + 4 = -\frac{21}{5} + 4$$

$$(p-2)^2 = -\frac{1}{5}$$

We cannot use the square root property to solve this equation, because the square root of $-\frac{1}{5}$ is not a real number. This equation has no real number solution. The solution set is $\varnothing$.

Objective 2 Practice Exercises

For extra help, see Examples 4–6 on pages 657–659 of your text.

Solve each equation by completing the square.

4. $6x^2 - x = 15$ 4. _____

5. $3x^2 - 2x + 4 = 0$ 5. _____

6. $3t^2 + t - 2 = 0$ 6. _____

Name: _____   Date: _____
Instructor: _____   Section: _____

Objective 3   Simplify the terms of an equation before solving.

Video Examples

Review this example for Objective 3:

7.   Solve $(x-6)(x+2)=7$.

$(x-6)(x+2)=7$

$x^2-4x-12=7$ FOIL

$x^2-4x=19$ Add 12.

$x^2-4x+4=19+4$ Add $\left[\frac{1}{2}(-4)\right]^2=4$

$(x-2)^2=23$

$x-2=\sqrt{23}$ or $x-2=-\sqrt{23}$

$x=2+\sqrt{23}$ or $x=2-\sqrt{23}$

The solution set is $\left\{2+\sqrt{23},\ 2-\sqrt{23}\right\}$.

Now Try:

7.   Solve $(x+8)(x-2)=5$.

_____

Objective 3 Practice Exercises

For extra help, see Examples 7–8 on page 659–670 of your text.

Simplify each of the following equations and then solve by completing the square.

7. $6y^2+3y=4y^2+y-5$

7. _____

8. $(b-1)(b+7)=9$

8. _____

9. $(s+3)(s+1)=1$ **9.** _____

Objective 4   Solve applied problems that require quadratic equations.

Video Examples

Review this example for Objective 4:

9. If James throws an object upward from ground level with an initial velocity of 80 feet per second, its height s (in feet) after t seconds is given by the formula $s = -16t^2 + 80t$. After how many seconds will the object reach a height of 64 feet?

Since s represents the height, we substitute 64 for s in the formula, and then solve for t using completing the square.

$$64 = -16t^2 + 80t$$

$$-4 = t^2 - 5t \qquad \text{Divide by } -16.$$

$$t^2 - 5t = -4$$

$$t^2 - 5t + \frac{25}{4} = -4 + \frac{25}{4} \qquad \text{Add } \left[\frac{1}{2}(-5)\right]^2 = \frac{25}{4}.$$

$$\left(t - \frac{5}{2}\right)^2 = \frac{9}{4}$$

$$t - \frac{5}{2} = \frac{3}{2} \quad \text{or} \quad t - \frac{5}{2} = -\frac{3}{2}$$

$$t = 4 \quad \text{or} \qquad t = 1$$

The ball reaches a height of 64 twice, once on the way up and again on the way down. It takes 1 second to reach 64 ft on the way up and then after 4 sec, the object reaches 64 ft on the way down.

Now Try:

9. A certain projectile is located at a distance of $d = 3t^2 - 6t + 1$ feet from its starting point after t seconds. How many seconds will it take the projectile to travel 10 feet?

_____

Copyright © 2018 Pearson Education, Inc.

Name: Date:
Instructor: Section:

Objective 4 Practice Exercises

For extra help, see Example 9 on page 660 of your text.

Solve.

10. A rule for estimating the number of board feet of
    lumber that can be cut from a log depends on the
    diameter and length of the log. To find the diameter
    d (in inches) needed to get x board feet of lumber
    from an 8-foot log, use the formula, $\left(\dfrac{d-4}{4}\right)^2 = x$.

    Find the diameter needed to get 20 board feet of
    lumber.

10. _____

11. The commodities market is very unstable; money
    can be made or lost quickly on investments in
    soybeans, wheat, pork bellies, and so on. Suppose
    that an investor kept track of his total profit, P (in
    thousands of dollars), at time t (in months), after he
    began investing, and found that his profit was given
    by the formula $P = 4t^2 - 24t + 32$. Find the times at
    which he broke even on his investment.

11. _____

12. George and Albert have found that the profit (in dollars) from their cigar shop is given by the formula $P = -10x^2 + 100x + 300$, where x is the number of units of cigars sold daily. How many units should be sold for a profit of \$460?

12. _____

Chapter 9 QUADRATIC EQUATIONS

9.3   Solving Quadratic Equations by the Quadratic Formula

Learning Objectives
1   Identify the values of a, b, and c in a quadratic equation.
2   Use the quadratic formula to solve quadratic equations.
3   Solve quadratic equations with one distinct solution.
4   Solve quadratic equations with fractional coefficients.

Key Terms

Use the vocabulary terms listed below to complete each statement in exercises 1–3.

quadratic formula   **standard form**   **constant**

1.   A quadratic equation written in the form $ax^2 + bx + c = 0$, $a \neq 0$ is written in

_____.

2.   A symbol that represents a value that doesn't change is a _____.

3.   The formula $x = \dfrac{-b \pm \sqrt{b^2 - 4ac}}{2a}$ is called the _____.

Objective 1   Identify the values of *a*, *b*, and *c* in a quadratic equation.

Video Examples

Review these examples for Objective 1:

1.   Identify the values of the variables a, b, and c in each quadratic equation $ax^2 + bx + c = 0$.

    a. $7x^2 - 6x + 4 = 0$

    Here $a = 7$, $b = -6$, and $c = 4$.

    b. $-8x^2 + 3 = 5x$

    Write the equation in standard form.
    $-8x^2 - 5x + 3 = 0$
    Here, $a = -8$, $b = -5$, and $c = 3$.

    e. $(3x - 5)(x + 6) = -29$

    $(3x - 5)(x + 6) = -29$

    $3x^2 + 13x - 30 = -29$

    $3x^2 + 13x - 1 = 0$
    Here, $a = 3$, $b = 13$, and $c = -1$.

Now Try:

1.   Identify the values of the variables a, b, and c in each quadratic equation $ax^2 + bx + c = 0$.

    a. $12x^2 - 10x + 7 = 0$

    _____

    b. $-4x^2 + 8 = 110x$

    _____

    e. $(5x - 4)(6x + 7) = 11$

    _____

Objective 1 Practice Exercises

For extra help, see Example 1 on page 663 of your text.

Write each equation in standard form, if necessary, and then identify the values of a, b, and c. Do not actually solve the equation.

1. $10x^2 = -4x$

1. _____

2. $4p = -4p^2 + 7$

2. _____

3. $(z+1)(z+2) = -7$

3. _____

Objective 2   Use the quadratic formula to solve quadratic equations.

Video Examples

Review these examples for Objective 2:	**Now Try:**

2.   Solve $2x^2 + 3x - 20 = 0$.

2.   Solve $3x^2 - x - 14 = 0$.

Here, $a = 2$, $b = 3$, and $c = -20$. Substitute into the quadratic formula.

_____

$$x = \frac{-b \pm \sqrt{b^2 - 4ac}}{2a}$$

$$x = \frac{-3 \pm \sqrt{3^2 - 4(2)(-20)}}{2(2)}$$

$$x = \frac{-3 \pm \sqrt{9 + 160}}{4}$$

$$x = \frac{-3 \pm \sqrt{169}}{4}$$

$$x = \frac{-3 \pm 13}{4}$$

Find the two solutions by first using the plus symbol, and then using the minus symbol.

$$x = \frac{-3 + 13}{4} = \frac{10}{4} = \frac{5}{2} \text{ or}$$

$$x = \frac{-3 - 13}{4} = \frac{-16}{4} = -4$$

Check each solution in the original equation. The solution set is $\left\{-4, \frac{5}{2}\right\}$.

Name: Date:
Instructor: Section:

3. Solve $x^2 = 4x - 1$.

Write the equation in standard form as
$x^2 - 4x + 1 = 0$.

$$x = \frac{-b \pm \sqrt{b^2 - 4ac}}{2a}$$

$$x = \frac{-(-4) \pm \sqrt{(-4)^2 - 4(1)(1)}}{2(1)}$$

$$x = \frac{4 \pm \sqrt{16 - 4}}{2}$$

$$x = \frac{4 \pm \sqrt{12}}{2}$$

$$x = \frac{4 \pm 2\sqrt{3}}{2}$$

$$x = \frac{2(2 \pm \sqrt{3})}{2}$$

$$x = 2 \pm \sqrt{3}$$

The solution set is $\{2 + \sqrt{3},\ 2 - \sqrt{3}\}$.

3. Solve $x^2 = 6x - 4$.

_____

Objective 2 Practice Exercises

For extra help, see Examples 2–4 on pages 665–666 of your text.

Use the quadratic formula to solve each equation. Write all radicals in simplified form, and write all answers in lowest terms.

4. $y^2 = 13 - 12y$

4. _____

5. $-7r^2 = 5r + 3$

5. _____

6. $5k^2 + 4k - 2 = 0$

6. _____

Objective 3   **Solve quadratic equations with one distinct solution.**

Video Examples

Review this example for Objective 3:

5.   Solve $9x^2 + 49 = 42x$.

Write the equation in standard form.
$$9x^2 - 42x + 49 = 0$$
Here, $a = 9$, $b = -42$, and $c = 49$.
$$x = \frac{-(-42) \pm \sqrt{(-42)^2 - 4(9)(49)}}{2(9)} = \frac{42}{18} = \frac{7}{3}$$
In this case, $b^2 - 4ac = 0$, and the trinomial
$9x^2 - 42x + 49$ is a perfect square. There is one
distinct solution in the solution set $\left\{\frac{7}{3}\right\}$.

Now Try:

5.   Solve $4x^2 - 44x + 121 = 0$.

_____

Objective 3 Practice Exercises

For extra help, see Example 5 on page 666 of your text.

Use the quadratic formula to solve each equation. Write all radicals in simplified form, and write all answers in lowest terms.

7. $m^2 + 4 = 4m$

7. _____

8. $4q^2 + 12q + 9 = 0$ **8.** _____

9. $49a^2 - 126a = -81$ **9.** _____

Objective 4   Solve quadratic equations with fractional coefficients.

Video Examples

Review this example for Objective 4:

6.   Solve $\frac{1}{5}t^2 = \frac{1}{5}t - \frac{3}{10}$.

Clear fractions. Multiply by the LCD, 10.

$$\frac{1}{5}t^2 = \frac{1}{5}t - \frac{3}{10}$$

$$10\left(\frac{1}{5}t^2\right) = 10\left(\frac{1}{5}t - \frac{3}{10}\right)$$

$$10\left(\frac{1}{5}t^2\right) = 10\left(\frac{1}{5}t\right) - 10\left(\frac{3}{10}\right)$$

$$2t^2 = 2t - 3$$

$2t^2 - 2t + 3 = 0$

Identify $a = 2$, $b = -2$, and $c = 3$.

$$t = \frac{-(-2) \pm \sqrt{(-2)^2 - 4(2)(3)}}{2(2)}$$

$$t = \frac{2 \pm \sqrt{4 - 24}}{4}$$

$$t = \frac{2 \pm \sqrt{-20}}{4}$$

Because $\sqrt{-20}$ does not represent a real number, the solution set is $\varnothing$.

Now Try:

6.   Solve $\frac{5}{7}x^2 = \frac{9}{7}x - \frac{12}{7}$.

_____

Name: Date:
Instructor: Section:

Objective 4 Practice Exercises

For extra help, see Example 6 on page 667 of your text.

Use the quadratic formula to solve each equation. Write all radicals in simplified form, and write all answers in lowest terms.

10. $\frac{1}{6}y^2 + \frac{1}{2}y = \frac{2}{3}$ 10. _____

11. $\frac{1}{4}t^2 - \frac{1}{3}t + \frac{5}{12} = 0$ 11. _____

12. $-\frac{1}{4}x^2 + 4 = \frac{1}{2}x$ 12. _____

Copyright © 2018 Pearson Education, Inc.

Chapter 9 QUADRATIC EQUATIONS

9.4   Graphing Quadratic Equations

Learning Objectives
1   Graph quadratic equations.
2   Find the vertex of a parabola.

Key Terms

Use the vocabulary terms listed below to complete each statement in exercises 1−4.

> **parabola   vertex   axis   line of symmetry**

1.   If a graph is folded on its_____, the two sides coincide.

2.   The _____ of a parabola that opens upward or downward is the lowest or highest point on the graph.

3.   The _____ of a parabola that opens upward or downward is a vertical line through the vertex.

4.   The graph of the quadratic equation $y = ax^2 + bx + c$ is called a

     _____.

Objective 1   Graph quadratic equations.

Video Examples

Review this example for Objective 1:

2.   Graph $y = x^2 - 3$.

     Find several ordered pairs. Let $x = 0$ to find the y-intercept.

     $$y = x^2 - 3 = 0^2 - 3 = -3$$

     This gives the ordered pair (0, –3). Select several values for x and find the corresponding values for y. Plot the ordered pairs and join them with a smooth curve.

Now Try:

2.   Graph $y = 9 - x^2$.

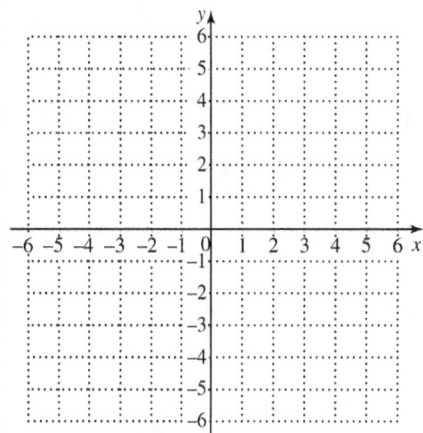

x	y
2	1
1	-2
0	-3
-1	-2
-2	1

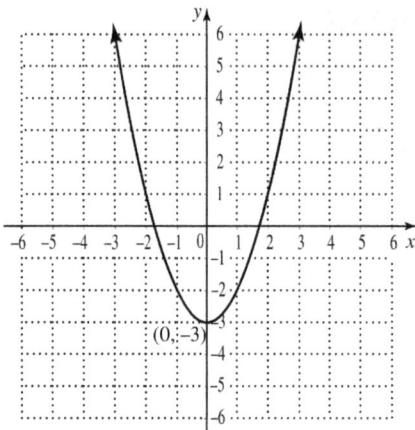

Objective 2   Find the vertex of a parabola.

Video Examples

Review these examples for Objective 2:

3. Graph $y = x^2 + x - 2$.

Find the x-intercepts of the parabola. Let $y = 0$.

$$0 = x^2 + x - 2$$
$$0 = (x+2)(x-1)$$
$$x + 2 = 0 \quad \text{or} \quad x - 1 = 0$$
$$x = -2 \quad \text{or} \quad x = 1$$

There are two intercepts, $(-2, 0)$ and $(1, 0)$. Since the x-value of the vertex is halfway between the x-values of the x-intercepts, it is half their sum.

$$x = \frac{1}{2}(-2 + 1) = -\frac{1}{2}$$

We find the corresponding y-value.

$$y = \left(-\frac{1}{2}\right)^2 + \left(-\frac{1}{2}\right) - 2 = -\frac{9}{4}$$

The vertex is $\left(-\frac{1}{2}, -\frac{9}{4}\right)$. The axis of symmetry

is the vertical line $x = -\frac{1}{2}$.

To find the y-intercept, we substitute 0 for x in the equation.

$$y = 0^2 + 0 - 2 = -2$$

We plot the three intercepts and the vertex, and find additional ordered pairs as needed.

Now Try:

3. Graph $y = x^2 - 2x - 3$.

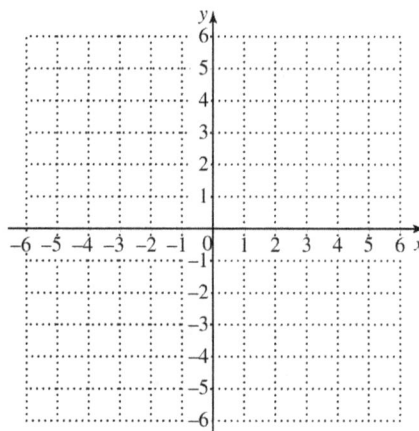

x	y
-2	0
-1	-2
$-\dfrac{1}{2}$	$-\dfrac{9}{4}$
0	-2
1	0

(graph with vertex labeled $(-1/2, -9/4)$)

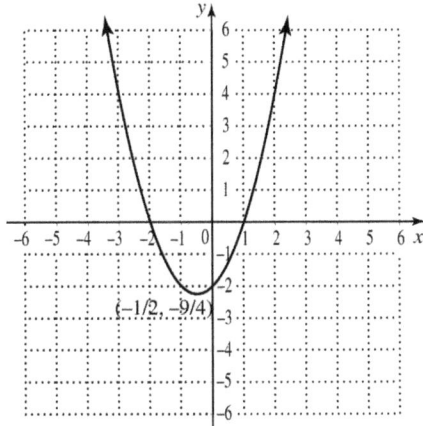

4. Graph $y = -x^2 + 4x - 1$.

Here $a = -1$ and $b = 4$, so we substitute to find the x-value of the vertex.

$$x = -\frac{b}{2a} = -\frac{4}{2(-1)} = 2$$

We find the y-value from the original equation.

$$y = -(2)^2 + 4(2) - 1$$
$$y = 3$$

The vertex is (2, 3). The axis is the line $x = 2$. Now we find the intercepts. Let $x = 0$ in the equation.

$$y = -0^2 + 4(0) - 1 = -1$$

The y-intercept is (0, –1). Let $y = 0$ to get the x-intercepts. We use the quadratic formula to solve for x.

$$x = \frac{-4 \pm \sqrt{(4)^2 - 4(-1)(-1)}}{2(-1)}$$

$$x = \frac{-4 \pm \sqrt{12}}{-2}$$

$$x = \frac{-4 \pm 2\sqrt{3}}{-2}$$

$$x = 2 \pm \sqrt{3}$$

Using a calculator, we find that the x-intercepts are (0.3, 0) and (3.7, 0).

4. Graph $y = x^2 + 8x + 14$.

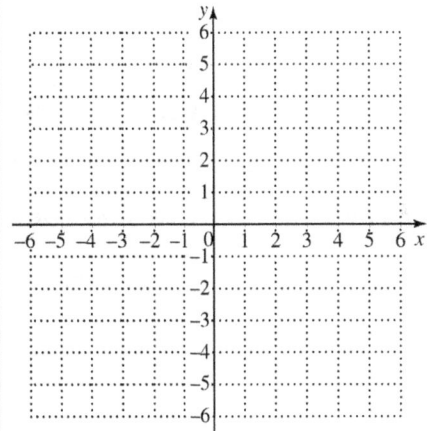

(blank graph grid)

We plot the intercepts, vertex, and other ordered pairs, and join these points with a smooth curve.

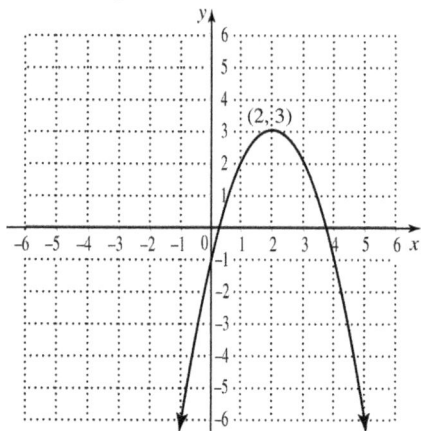

x	y
0	−1
0.3	0
1	2
2	3
3	2
3.7	0

Objective 1 & 2 Practice Exercises

For extra help, see Examples 1–4 on pages 672–675 of your text.

Graph each equation. Give the coordinates of the vertex in each case.

1. $y = -x^2 - 1$

1.

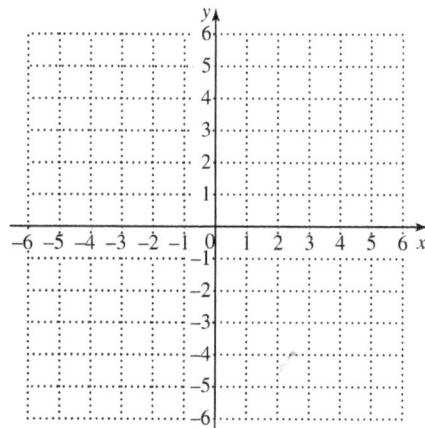

vertex: _____

2. $y = (x-2)^2$

2.

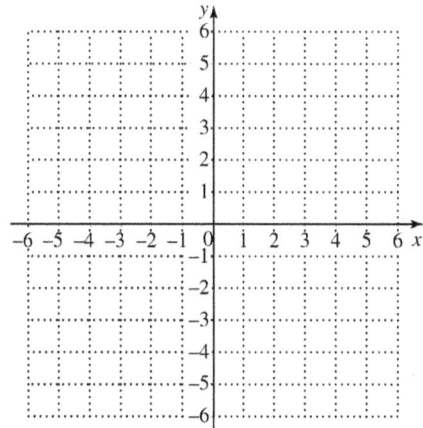

vertex: _____

3. $y = 2x^2 + 4x$

3.

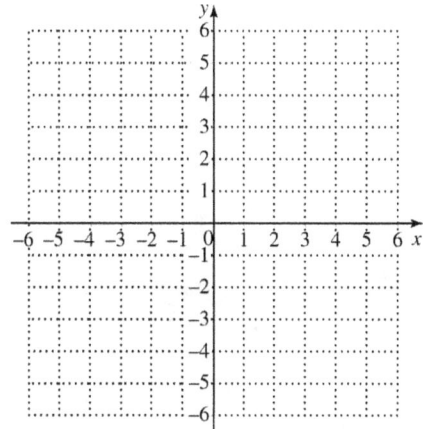

vertex: _____

4. $y = -x^2 - 3x + 1$

4.

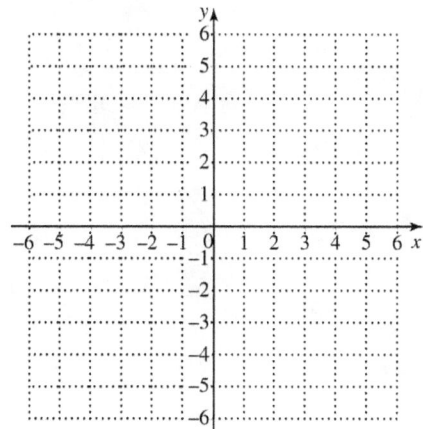

vertex: _____

5. $y = -\dfrac{1}{2}x^2 + 2x$

5.

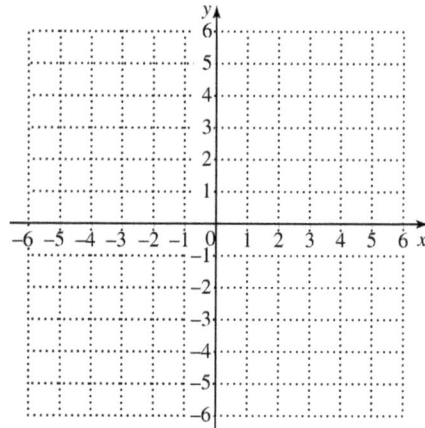

vertex: _____

6. $y = \dfrac{1}{3}x^2$.

6.

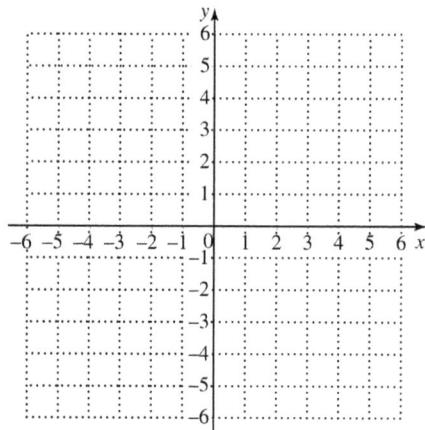

vertex: _____

Chapter 9 QUADRATIC EQUATIONS

9.5   Introduction to Functions

Learning Objectives	
1	Understand the definition of a relation.
2	Understand the definition of a function.
3	Determine whether a graph or equation represents a function.
4	Use function notation.
5	Apply the function concept in an application.

Key Terms

Use the vocabulary terms listed below to complete each statement in exercises 1–5.

components	relation	domain	range	function

1. Any set of ordered pairs is called a _____.

2. The set of all second components in the ordered pairs of a relation is the _____ of the relation.

3. A _____ is a set of ordered pairs in which each first component corresponds to exactly one second component.

4. In an ordered pair (x, y), x and y are the _____.

5. The set of all first components in the ordered pairs of a relation is the _____ of the relation.

Objective 1   Understand the definition of a relation.

Video Examples

Review these examples for Objective 1:

1. Identify the domain and range of each relation.

   a. $\{(1, 0), (2, 7), (5, 9), (6, 2)\}$

   This relation has
      domain: $\{1, 2, 5, 6\}$
   and
      range: $\{0, 7, 9, 2\}$

   b. $\{(5, 7), (5, -10), (5, 8), (5, 4)\}$

   This relation has
      domain: $\{5\}$
   and
      range: $\{7, -10, 8, 4\}$

Now Try:

1. Identify the domain and range of each relation.

   a. $\{(6, 7), (8, 9), (10, 11), (12, 13)\}$

   _____

   b. $\{(1, 11), (1, 12), (1, 13), (1, 14)\}$

   _____

Objective 1 Practice Exercises

For extra help, see Example 1 on page 678 of your text.

Identify the domain and range of each relation.

1. $\{(2,7),(5,-4),(-3,-1),(0,-8),(5,2)\}$ **1.**

                 domain:_____

                 range: _____

2. $\{(3,5),(3,8),(3,-4),(3,1),(3,0)\}$ **2.**

                 domain:_____

                 range: _____

3. $\{(-3,5),(-2,5),(-1,0),(0,-5),(1,5)\}$ **3.**

                 domain:_____

                 range: _____

Objective 2   Understand the definition of a function.

Video Examples

Review these examples for Objective 2:

2. Determine whether each relation is a function.

   a. {(–5, 10), (–4, 8), (0, 0), (4, 8), (5, 10)}

   Each first component appears once and only once. The relation is a function.

   b. {(7, 4), (–7, 3), (7, 2)}

   The first component 7 appears in two ordered pairs and corresponds to two different second components. Therefore, this relation is not a function.

Now Try:

2. Determine whether each relation is a function.

   a. {(0, 1), (4, 2), (5, 9), (6, 15)}

      _____

   b. {(10, 0), (10, 5), (10, 20)}

      _____

Objective 2 Practice Exercises

For extra help, see Example 2 on page 679 of your text.

Determine whether each relation is a function.

4. $\{(1,3),(5,7),(11,9),(8,-2),(6,-7),(-4,-3)\}$ 4. _____

5. $\{(-1.2,4),(1.8,-2.5),(3.7,-3.8),(3.7,3.8)\}$ 5. _____

6. $\{(-3,5),(-2,5),(-1,0),(0,-5),(1,5)\}$ 6. _____

Objective 3   Determine whether a graph or equation represents a function.

Video Examples

Review these examples for Objective 3:

3. Determine whether each relation represented by a graph or an equation is a function.

c.

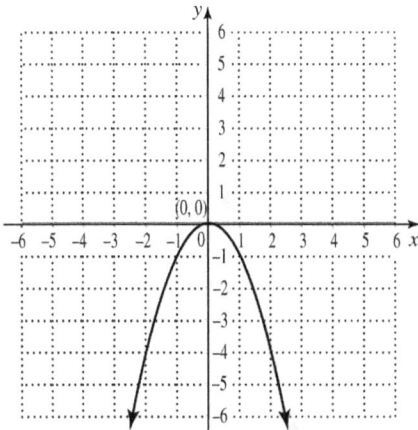

Apply the vertical line test. Any vertical line intersects the graph of a vertical parabola just once, so this relation is a function.

Now Try:

3. Determine whether each relation represented by a graph or an equation is a function.

c.

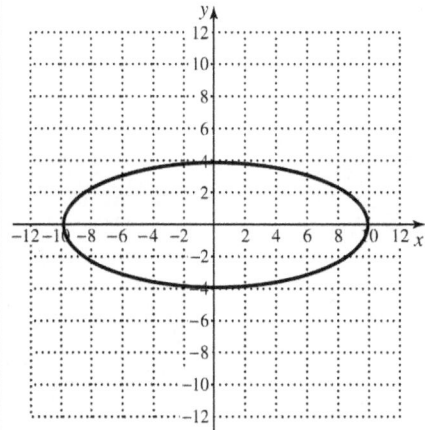

_____

d.

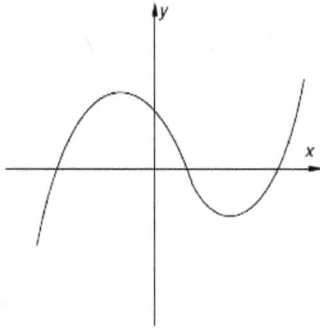

d.

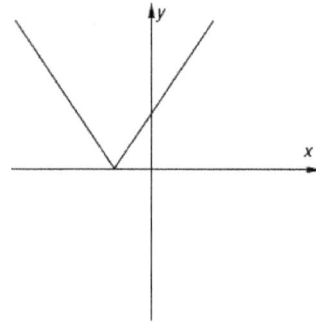

Use the vertical line test. Any vertical line intersects the graph just once, so this is the graph of a function.

_____

Objective 3 Practice Exercises

For extra help, see Example 3 on pages 680–681 of your text.

Use the vertical line test to determine whether each relation graphed is a function.

7.

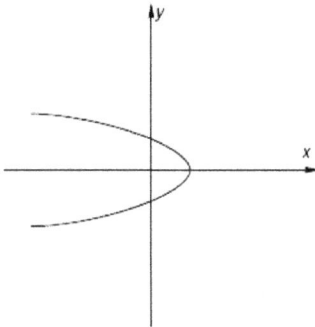

7. _____

8.

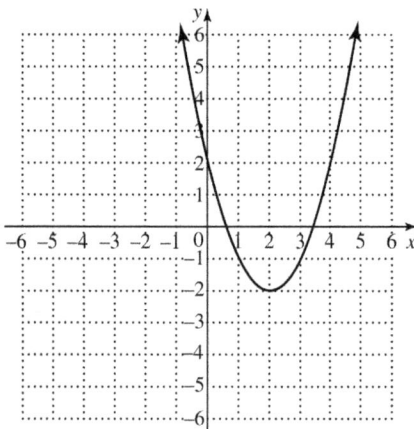

8. _____

Decide whether the equation defines y as a function of x.

9. $y = 7$

9. _____

Objective 4   Use function notation.

Video Examples

Review these examples for Objective 4:	Now Try:

Review these examples for Objective 4:

4. For the function $f(x) = x^2 - 5$, find each function value.

    a. $f(3)$

    Substitute 3 for x.

$$f(x) = x^2 - 5$$
$$f(3) = 3^2 - 5$$
$$f(3) = 9 - 5$$
$$f(3) = 4$$

    b. $f(0)$

$$f(0) = 0^2 - 5$$
$$f(0) = 0 - 5$$
$$f(0) = -5$$

    c. $f(-4)$

$$f(-4) = (-4)^2 - 5$$
$$f(-4) = 16 - 5$$
$$f(-4) = 11$$

Now Try:

4. For the function $f(x) = 5x - 10$, find each function value.

    a. $f(3)$

    _____

    b. $f(0)$

    _____

    c. $f(-1)$

    _____

Objective 4 Practice Exercises

For extra help, see Example 4 on page 682 of your text.

For each function f, find (a) $f(-2)$, (b) $f(0)$, *and* (c) $f(4)$.

10. $f(x) = 3x - 7$

    10. a. _____

    b. _____

    c. _____

11. $f(x) = x^2 + 2$ 11. a._____

                                            b._____

                                            c._____

12. $f(x) = 9$ 12. a._____

                                            b._____

                                            c._____

Objective 5   Apply the function concept in an application.

Video Examples

Review these examples for Objective 5:
5.

Year	Profit (millions of dollars)
2006	34
2008	40
2010	46
2012	48

a. Write the information in the graph as a set of ordered pairs. Does this set define a function?

{(2006, 34), (2008, 40), (2010, 46), (2012, 48)}
Yes, this set defines a function.

b. Suppose that p is the name given to this relation. Give the domain and range of p.

The domain of the function is the set of years.
    Domain: {2006, 2008, 2010, 2012}
The range of the function is the set of profits, in millions.
Range: {34, 40, 46, 48}

Now Try:
5.

Worldwide Internet Users

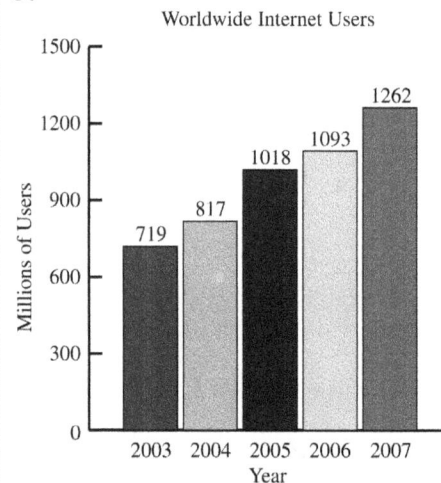

Source: www. internetworldstats.com-January, 2008

a. Write the information in the graph as a set of ordered pairs. Does this set define a function?

_____

b. Suppose that w is the name given to this relation. Give the domain and range of w.

_____

_____

c. Find $p(2008)$ and $p(2012)$.

We refer to the table or ordered pairs from (a).
$p(2008) = 40$ million and $p(2012) = 48$ million

d. For what value of x does $p(x) = 46$ (million)?

We use the table or the ordered pairs from (a).
$p(2010) = 46$ million

c. Find $w(2003)$ and $w(2006)$.

_____

d. For what value of x does $w(x) = 1018$ (million)?

_____

Objective 5 Practice Exercises

For extra help, see Example 5 on page 682 of your text.

The yearly revenue for a small business is shown in the table below. Use the table to answer problems 13−15.

Year	Revenue (thousands of dollars)
2008	596
2009	625
2010	872
2011	795
2012	625

13.   Write the information in the table as a set of ordered pairs. Does this set define a function?

13. _____

14.   Suppose that r is the name given to this relation. Give the domain and range of r.

14. domain _____

     range_____

15.   Find $r(2009)$ and $r(2011)$.

15. _____

Chapter R PREALGEBRA REVIEW

R.1   Fractions

Key Terms

1. equivalent fractions
2. improper fraction
3. numerator
4. proper fraction
5. denominator
6. composite number
7. prime factorization
8. prime number
9. lowest terms

Objective 1
  Practice Exercises
  1. prime
  3. neither

Objective 2
  Now Try
  2a. $5 \cdot 11$
  2b. $2 \cdot 3 \cdot 5 \cdot 7$

  Practice Exercises
  5. $2 \cdot 2 \cdot 2 \cdot 2 \cdot 2 \cdot 2 \cdot 2 \cdot 2$

Objective 3
  Now Try
  3c. $\dfrac{3}{4}$

  Practice Exercises
  7. $\dfrac{7}{25}$
  9. $\dfrac{33}{73}$

Objective 4
  Now Try
  4. $14\dfrac{4}{5}$
  5. $\dfrac{86}{7}$

  Practice Exercises
  11. $\dfrac{122}{9}$

Objective 5
  Now Try
  7b. $\dfrac{16}{21}$
  7c. $\dfrac{1}{10}$

  Practice Exercises
  13. $\dfrac{15}{2}$ or $7\dfrac{1}{2}$
  15. $\dfrac{45}{4}$ or $11\dfrac{1}{4}$

Objective 6
  Now Try
  9a. $\dfrac{23}{24}$
  10b. $\dfrac{41}{42}$

Practice Exercises

17. $\frac{125}{12}$ or $10\frac{5}{12}$

Objective 7
Now Try

12. $14\frac{2}{9}$ pies

Practice Exercises

19. $220.50 $3\frac{1}{4}$ yards   21. $34\frac{5}{8}$ yards

Objective 8
Now Try

13c. 104 workers

Practice Exercises

23. About 780 workers

R.2   Decimals and Percents

Key Terms

1. decimals   2. place value   3. percent

Objective 1
Now Try

1c. $\frac{37,058}{10,000}$

Practice Exercises

1. $\frac{7}{1000}$ 3. $\frac{300,005}{10,000}$

Objective 2
Now Try

2c. 7.024

Practice Exercises

5. 755.098

Objective 3
Now Try

3a. 251.116   4a. 32.4   5a. 5130.2

5b. 0.0986

Practice Exercises

7. 2.3424   9. 0.4292

Objective 4
Now Try
6a. 0.35   6b. $3.\overline{5}$; 3.556

Practice Exercises
11. $0.\overline{4}$ or 0.444

Objective 5
Now Try
9c. 0.0043   9d. 91%

Practice Exercises
13. 3.62   15. 0.84%

Objective 6
Now Try
10a. $\dfrac{13}{25}$

Practice Exercises
17. $\dfrac{7}{200}$

Objective 7
Now Try
12. $14, $42 saved

Practice Exercises
19. $810   21. 20%

Chapter 1 THE REAL NUMBER SYSTEM

1.1   Exponents, Order of Operations, and Inequality

Key Terms
1. exponential expression   2. base

3. inequality   4. exponent

Objective 1
  Now Try
1a. 81

  Practice Exercises
1. 27   3. 0.16

Objective 2
  Now Try
2a. 85

  Practice Exercises
5. 45

Objective 3
  Now Try
3a. 215

  Practice Exercises
7. −8   9. 48

Objective 4
  Now Try
4c. true

  Practice Exercises
11. false

Objective 5
  Now Try
5d. $5 > 3$

  Practice Exercises
13. $7 = 13 - 6$ 15. $20 \geq 2 \cdot 7$

Objective 6
  Now Try
6a. $11 < 15$

  Practice Exercises
17. $8 \leq 12$

1.2   Variables, Expressions, and Equations

Key Terms
1. equation
2. variable
3. algebraic expression
4. solution
5. constant

Objective 1
Now Try
1d. 252, 567   2a. 64   2b. 20

Practice Exercises
1. 8   3. $\dfrac{28}{13}$

Objective 2
Now Try
3c. $12 - x$ 3f. $5(x-6)$

Practice Exercises
5. $8x - 11$

Objective 3
Now Try
4a. yes   4b. no

Practice Exercises
7. no   9. yes

Objective 4
Now Try
5c. $3x - 7 = 12$

Practice Exercises
11. $6(5+x) = 19$

Objective 5
Now Try
6a. equation   6b. expression

Practice Exercises
13. expression   15. equation

1.3   Real Numbers and the Number Line

Key Terms
1. whole numbers
2. additive inverse
3. integers
4. natural numbers
5. absolute value
6. number line
7. irrational number
8. coordinate
9. negative number
10. positive number
11. real numbers
12. set-builder notation
13. signed numbers
14. rational number

Answers

Objective 1
Now Try
1a. -282

2.

3a. 7   3b. 0, 7   3c. -10, 0, 7

3d. $-10, -\frac{5}{8}, 0, 0.\overline{4}, 5\frac{1}{2}, 7, 9.9$ 3e. $\sqrt{5}$

3f. all of the numbers

Practice Exercises
1. -75 pounds   3.

Objective 2
Now Try
4. true

Practice Exercises
5. false

Objective 3
Practice Exercises
7. 25   9. -4.5

Objective 4
Now Try
5b. 10   5c. 10   5e. -10

Practice Exercises
11. 1.22

1.4   Adding Real Numbers

Key Terms
1. sum   2. addends

Objective 1
Now Try
1a. 6   1b. -5

2a. -12 2b. -31 2c. -40

Practice Exercises
1. -18 3. $-5\frac{5}{8}$

Objective 2
Now Try
4a. -12 4b. -7

Practice Exercises

5. $-\dfrac{1}{6}$

Objective 3
   Now Try
5b. 15

   Practice Exercises
7. −3   9. −6.8

Objective 4
   Now Try
6a. $-10 + 11 + 2$; 3   6b. $(-9 + 15) + 6$; 12

   Practice Exercises
11. $-10 + \left[20 + (-4)\right]$; 6

1.5   Subtracting Real Numbers

Key Terms
   1. minuend   2. subtrahend   3. difference

Objective 1
   Now Try
   1. 2

   Practice Exercises
   1. 3   3. −7

Objective 2
   Now Try
2b. −4   2c. −38   2d. 2

2e. $\dfrac{61}{45}$

   Practice Exercises
   5. 4.4

Objective 3
   Now Try

3a. −5   3b. $\dfrac{19}{2}$

   Practice Exercises
   7. 18   9. $-\dfrac{23}{18}$

Objective 4
   Now Try
4a. $-17 - 9$; −26   4b. $\left[25 + (-6)\right] - 8$; 11   5. 5464 ft

Answers

11. $(-4+12)-9;\ -1$

1.6 Multiplying and Dividing Real Numbers

Key Terms
1. quotient   2. reciprocals   3. product

4. dividend   5. divisor

Objective 1
  Now Try
1a. -56 1c. -45.08

  Practice Exercises
1. -28 3. -13.12

Objective 2
  Now Try
2a. 30 2b. 135

  Practice Exercises
5. $\dfrac{4}{5}$

Objective 3
  Now Try
3a. 3 3b. -2 3c. 3

3d. $\dfrac{4}{11}$

  Practice Exercises
7. 6 9. undefined

Objective 4
  Now Try
4d. $-\dfrac{5}{4}$

  Practice Exercises
11. $-\dfrac{11}{2}$

Objective 5
  Now Try
5a. -432 5b. 103

  Practice Exercises
13. 7 15. $\dfrac{19}{4}$

     Copyright © 2018 Pearson Education, Inc.

Objective 6
  Now Try
  6a. $16[5+(-7)];\ -32$ 6d. $0.08[18-(-4)];\ 1.76$
  Practice Exercises
  17. $85-\dfrac{3}{10}[50-(-10)];\ 67$

Objective 7
  Now Try
  8d. $\dfrac{36}{x}=-4$
  Practice Exercises
  19. $\dfrac{2}{3}x=-7$ 21. $\dfrac{x}{-4}=1$

1.7   Properties of Real Numbers

Key Terms
  1. identity element for addition
  2. identity element for multiplication

Objective 1
  Now Try
  1a. -12 1b. 2
  Practice Exercises
  1. 4 3. $(4+z)$

Objective 2
  Now Try
  2a. 4 2b. $[(-3)\cdot4]$ 4a. 107
  4b. $12{,}600$
  Practice Exercises
  5. $[(-4+3y)]$

Objective 3
  Now Try
  6a. $\dfrac{7}{9}$
  Practice Exercises
  7. 4 9. $\dfrac{6}{7}$

Objective 4
  Now Try
  7b. -8 7d. $\dfrac{5}{8}$

Answers

Practice Exercises
11. 0; identity

Objective 5
 Now Try
8f. 375   9c. $4x + 5y - z$

Practice Exercises
13. $2an - 4bn + 6cn$ 15. $2k - 7$

1.8   Simplifying Expressions

Key Terms
   1. numerical coefficient   2. term

   3. like terms

Objective 1
 Now Try
1c. $71 + 18x$ 1d. $11 - 7x$

  Practice Exercises
   1. $8x + 27$ 3. $10x + 3$

Objective 2
  Practice Exercises
   5. $\dfrac{7}{9}$

Objective 3
  Practice Exercises
   7. like   9. unlike

Objective 4
 Now Try
2c. $19x$ 3a. $49y + 15$

  Practice Exercises
  11. $2x - 14$

Objective 5
 Now Try
   4. $11 + 10x + 8x + 4x$; $11 + 22x$

  Practice Exercises
  13. $6x + 12 + 4x = 10x + 12$
  15. $4(2x - 6x) + 6(x + 9) = -10x + 54$

Chapter 2 EQUATIONS, INEQUALITIES, AND APPLICATIONS

2.1   The Addition Property of Equality

Key Terms

1. equivalent equations

2. linear equation

3. solution set

Objective 1
Practice Exercises

1. no

3. yes

Objective 2
Now Try

1. {21}

3. {−19}

4. {7}

Practice Exercises

5. $\left\{\dfrac{1}{2}\right\}$

Objective 3
Now Try

7. {29}

8. {8}

Practice Exercises

7. {7}

9. {7.2}

2.2   The Multiplication Property of Equality

Key Terms

1. multiplication property of equality

2. addition property of equality

Objective 1
Now Try

1. {14}

6. {39}

4. {24}

5. {36}

Practice Exercises

1. {−17}

3. {6.4}

Objective 2
Now Try

7. {4}

Practice Exercises

5. {−5}

2.3 More on Solving Linear Equations

Key Terms

1. contradiction   2. conditional equation   3. identity

Objective 1

Now Try

3. $\{4\}$

Practice Exercises

1. $\left\{\dfrac{5}{2}\right\}$ 3. $\left\{-\dfrac{1}{5}\right\}$

Objective 2

Now Try

6. {all real numbers}   7. $\varnothing$

Practice Exercises

5. {all real numbers}

Objective 3

Now Try

8. $\{-18\}$ 10. $\{2\}$

Practice Exercises

7. $\{2\}$ 9. $\{10\}$

Objective 4

Now Try

11a. $67 - t$

Practice Exercises

11. $\dfrac{17}{p}$

2.4 An Introduction to Applications of Linear Equations

Key Terms

1. supplementary angles   2. complementary angles

3. right angle   4. straight angle   5. consecutive integers

Objective 1

Practice Exercises

1. Read the problem; assign a variable to represent the unknown; write an equation; solve the equation; state the answer; check the answer.

Objective 2

Now Try

1. 16

Practice Exercises
3. $-2(4-x)=24$; 16

Objective 3
Now Try
2. 52

Practice Exercises
5. $x+(x+5910)=34,730$; Mt. Rainier: 14,410 ft; Mt. McKinley: 20,320 feet
7. $x+(5+3x)+4x=29$; Mark: 3 laps; Pablo: 14 laps; Faustino: 12 laps

Objective 4
Now Try
5. 352 and 353   6. $-2, 0, 2, 4$

Practice Exercises
9. 27, 28

Objective 5
Now Try
7. 18°

Practice Exercises
11. 133°   13. 27°

2.5   Formulas and Additional Applications from Geometry

Key Terms
1. vertical angles   2. formula   3. perimeter

4. area

Objective 1
Now Try
1a. $W=5.5$

Practice Exercises
1. $a=36$ 3. $h=12$

Objective 2
Now Try
2. 12 ft   3. 15 ft, 20 ft, 30 ft

Practice Exercises
5. 1.5 years

Objective 3
Now Try
5b. 54°, 126°

Answers

Practice Exercises

7. 35°, 35°

9. 129°, 51°

Objective 4
  Now Try

6. $t = \dfrac{d}{r}$

7. $a = P - b - c$

9b. $y = 6x + 5$

Practice Exercises

11. $n = \dfrac{S}{180} + 2$ or $n = \dfrac{S + 360}{180}$

2.6   Ratio, Proportion, and Percent

Key Terms

1. proportion

2. ratio

3. terms

4. cross products

Objective 1
  Now Try

1a. $\dfrac{11}{17}$

1b. $\dfrac{4}{15}$

2. 24-ounce jar, $0.054 per oz

Practice Exercises

1. $\dfrac{8}{3}$

3. 45-count box

Objective 2
  Now Try

4. $\left\{ -\dfrac{3}{4} \right\}$

Practice Exercises

5. $\left\{ \dfrac{10}{3} \right\}$

Objective 3
  Now Try

5. $259.20

Practice Exercises

7. 15 inches

9. $135

Objective 4
  Now Try

6a. 140

6b. 950

Practice Exercises

11. 2%

2.7   Solving Linear Inequalities

Key Terms

1. three-part inequality   2. interval

3. linear inequality   4. inequalities   5. interval notation

Objective 1
Now Try

1.

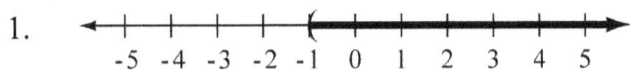

Practice Exercises

1. $(3, \infty)$;

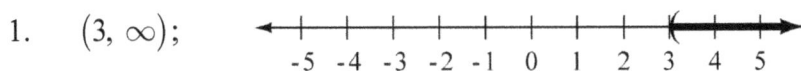

3. $(-\infty, -4)$;

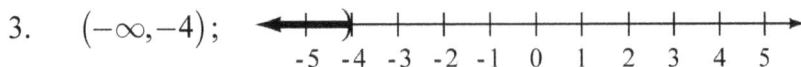

Objective 2
Now Try

3. $[-3, \infty)$

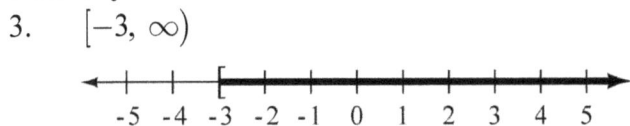

Practice Exercises

5. $(2, \infty)$;

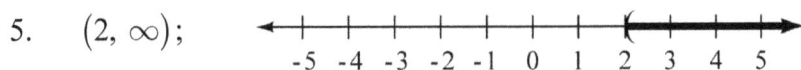

Objective 3
Now Try

4a. $(-\infty, -5]$

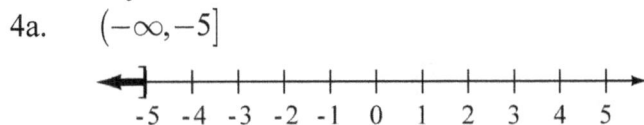

4b. $(-\infty, -4)$

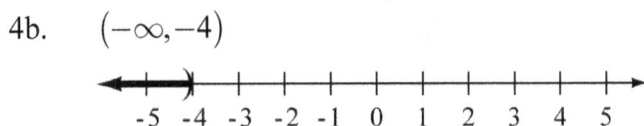

Practice Exercises

7. $(-2, \infty)$;

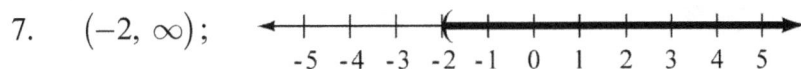

Answers

9. $(-\infty, 4]$

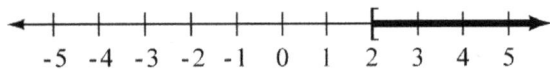

Objective 4
  Now Try
  5. $[2, \infty)$

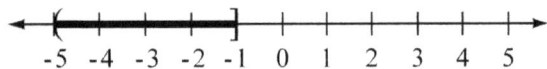

  Practice Exercises

11. $(-\infty, 5]$;

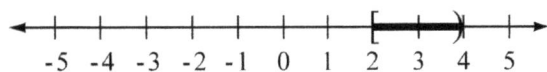

Objective 5
  Now Try
  8. 10 feet

  Practice Exercises
13. 89   15. all numbers greater than 5

Objective 6
  Now Try
  9. $(-5, -1]$

  10a. $[2, 4)$

  Practice Exercises

17. $[-5, -3)$;

Chapter 3 GRAPHS OF LINEAR EQUATIONS AND INEQUALITIES IN TWO VARIABLES

3.1   Linear Equations and Rectangular Coordinates

Key Terms

1. line graph
2. linear equation in two variables
3. coordinates
4. x-axis
5. y-axis
6. ordered pair
7. rectangular (Cartesian) coordinate system
8. quadrants
9. origin
10. plot
11. scatter diagram
12. table of values
13. plane

Objective 1

Now Try

1a. 2010-2011, 2012-2013, 2014-2015   1b. 2011-2012
1c. 300 degrees

Practice Exercises

1. 2013-2014   3. 1600 and 1000 degrees; 600 degrees

Objective 2

Practice Exercises

5. $\left(0, \frac{1}{3}\right)$

Objective 3

Now Try

2a. yes   2b. no

Practice Exercises

7. no, not a solution   9. yes, a solution

Objective 4

Now Try

3a. (5, 13)

Practice Exercises

11. (a) $(-4, -5)$; (b) $(2, 7)$; (c) $\left(-\frac{3}{2}, 0\right)$; (d) $(-2, -1)$; (e) $(-5, -7)$

Objective 5

Now Try

4a.

x	y
1	-4
5	12
2	0
3	4

$(1, -4), (5, 12), (2, 0), (3, 4)$

Answers

Practice Exercises

13. $(-4, 4), (0, 4), (6, 4)$

Objective 6
Now Try

5.

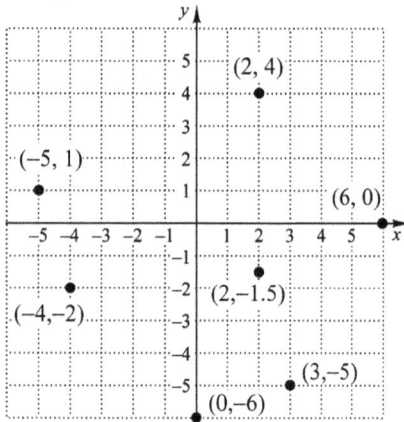

6a.

x (months)	y (balance due)
2	10,153.48
5	10,565.20
11	11,388.64

6b.

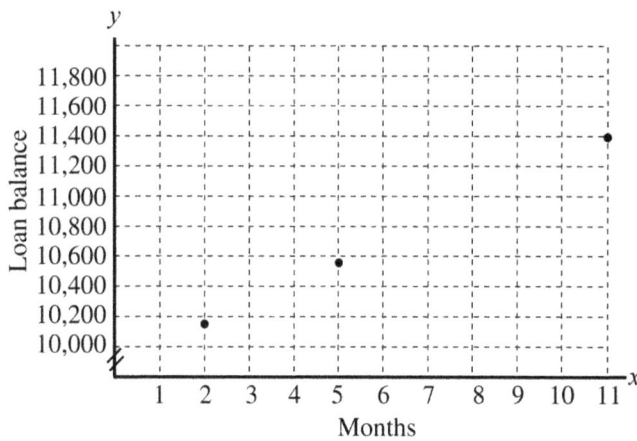

Practice Exercises

15.

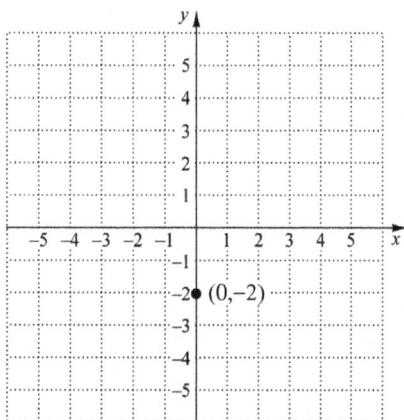

17.

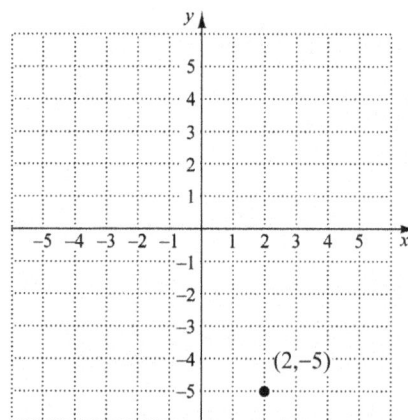

Copyright © 2018 Pearson Education, Inc.

3.2   Graphing Linear Equations in Two Variables

Key Terms
1. *y*-intercept   2. *x*-intercept   3. graphing
4. graph

Objective 1
Now Try
2.

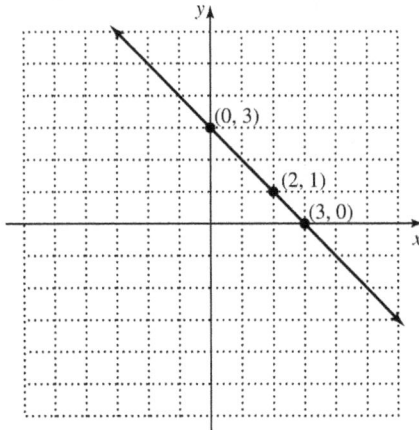

Practice Exercises

1. $(0, -2), \left(\frac{2}{3},\ 0\right), (2,\ 4)$

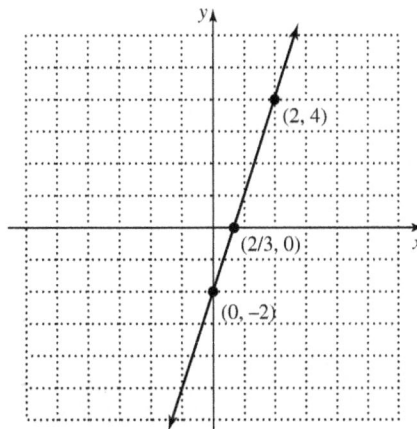

3. $\left(0, -\frac{1}{2}\right), (1,\ 0), (-3, -2)$

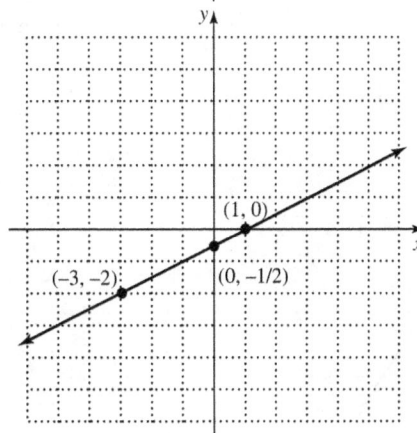

Answers

Objective 2
Now Try
3.

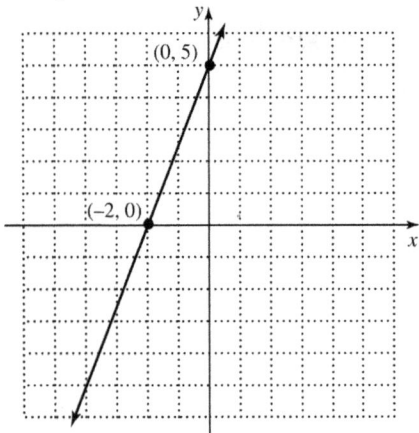

Practice Exercises
5.

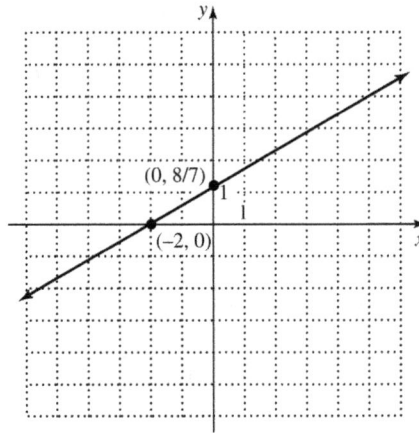

Objective 3
Now Try
5.

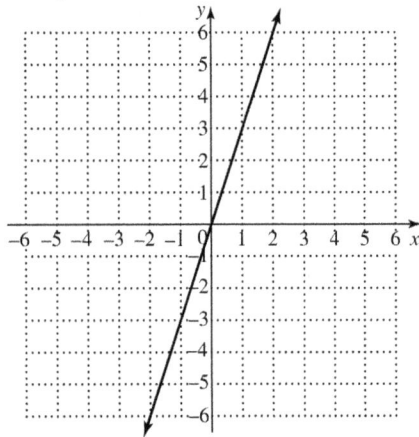

Practice Exercises
7. $-3x - 2y = 0$

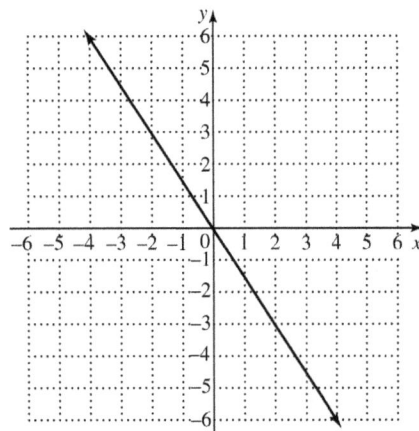

9. $y = 2x$

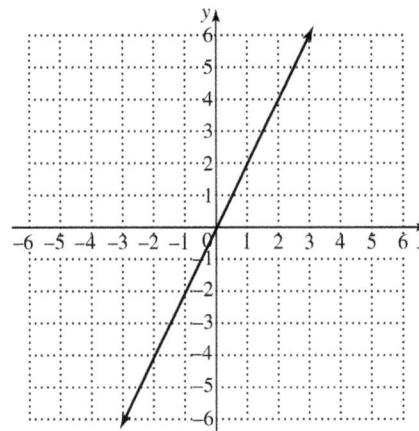

Copyright © 2018 Pearson Education, Inc.

Objective 4
Now Try

6.

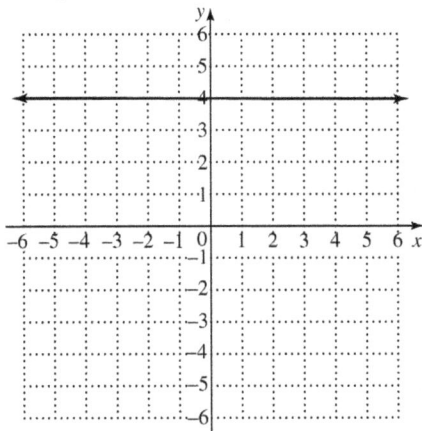

7.

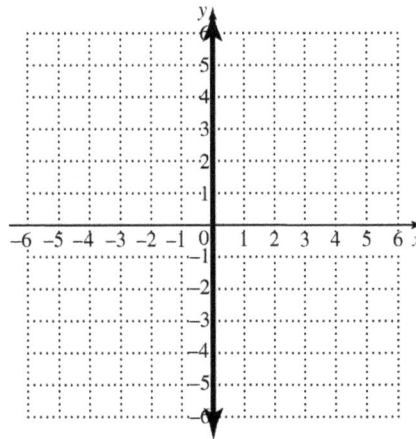

Practice Exercises

11. $y + 3 = 0$

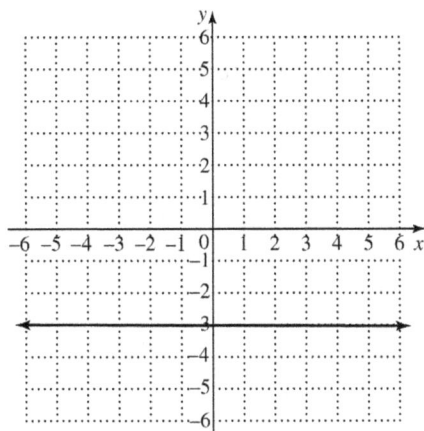

Objective 5
Now Try

8a. 0 calculators, $45
5000 calculators, $42
20,000 calculators, $33
45,000 calculators, $18

8b. (0, $45), (5, $42), (20, $33), (45, $18)

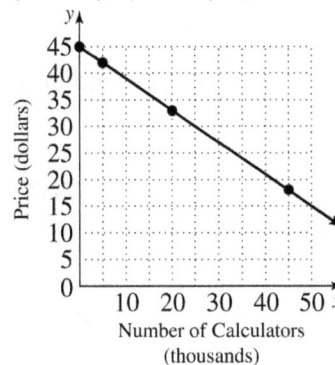

8c. 30,000 calculators, $27

Answers

Practice Exercises

13. 2013, 325;
    2014, 367;
    2015, 409;
    2016, 451

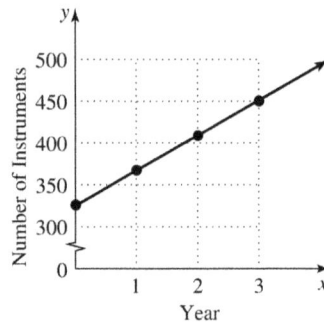

3.3   The Slope of a Line

Key Terms

1. perpendicular lines   2. slope   3. rise

4. parallel lines   5. run

Objective 1
Now Try

1. $\frac{4}{1}$, or 4   3. 0   4. undefined slope

2a. $-\frac{16}{9}$ 2b. $\frac{9}{17}$

Practice Exercises

1. -2 3. 0

Objective 2
Now Try

5a. $\frac{7}{4}$ 5b. $-\frac{3}{2}$ 5c. $-\frac{1}{8}$

Practice Exercises

5. $\frac{2}{3}$

Objective 3
Now Try

6b. parallel   6a. perpendicular

Practice Exercises

7. 1; 1; parallel   9. -3; $\frac{1}{3}$; perpendicular

3.4   Slope-Intercept Form of a Linear Equation

Key Terms

1. slope-intercept form   2. standard form   3. *y*-intercept

4. slope

Objective 1
Now Try
1a. slope: -12; y-intercept: $(0, 6)$

1c. slope: 23; y-intercept $(0, 0)$

Practice Exercises
1. slope: $\frac{3}{2}$; y-intercept: $\left(0, -\frac{2}{3}\right)$

3. $y = -3x + 3$

Objective 2
Now Try

2b.

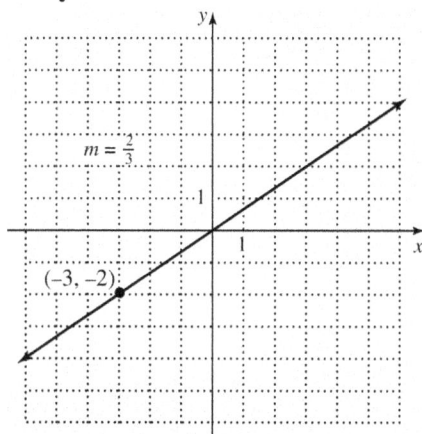

3.

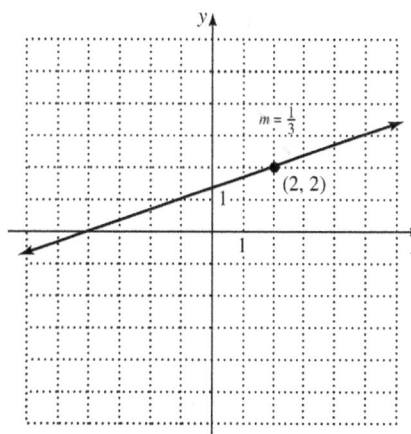

Practice Exercises

5.

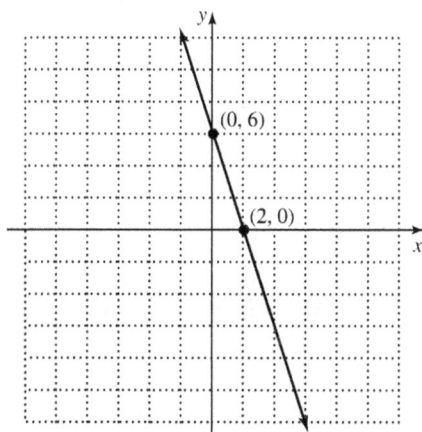

Objective 3
Now Try
4a. $y = \frac{3}{4}x - 6$

4b. $y = 6x + 10$

Practice Exercises
7. $y = -2x - 11$

Answers

Objective 4
Now Try
5a.

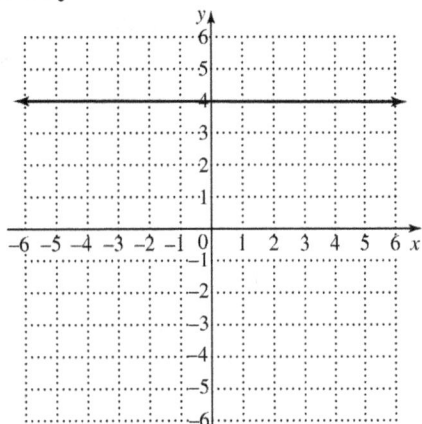

Practice Exercises
9.

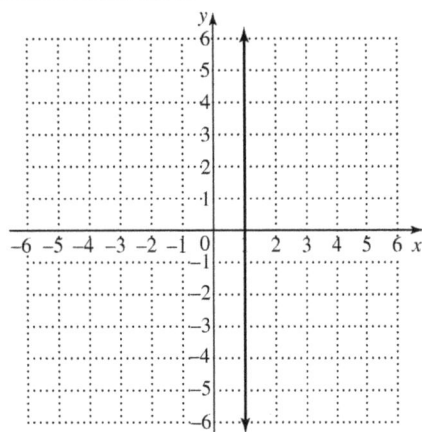

11. $x = 3$

3.5 Point-Slope Form of a Linear Equation and Modeling

Key Terms
1. point-slope form   2. standard form   3. slope-intercept form

Objective 1
Now Try
1b. $y = -\dfrac{2}{3}x + 15$

Practice Exercises
1. $y = \dfrac{1}{4}x - 9$ 3. $y = 2x - 8$

Objective 2
Now Try
2. $y = -\dfrac{3}{4}x + \dfrac{81}{4}$; $3x + 4y = 81$

Practice Exercises
5. $3x - 2y = 0$

    Copyright © 2018 Pearson Education, Inc.

Objective 3
Now Try

3.

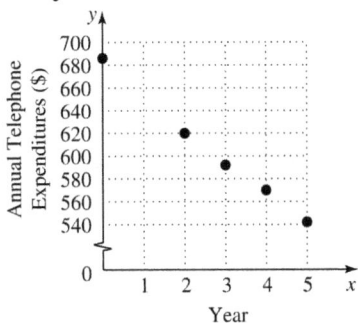

$$y = -\frac{144}{5}x + 686$$

Practice Exercises

7.

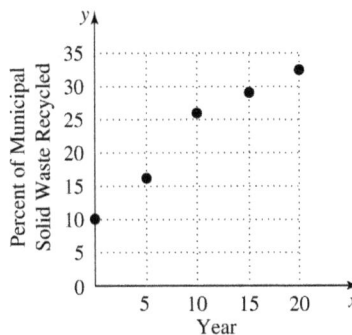

$$y = \frac{28}{25}x + 10.1$$

3.6　Graphing Linear Inequalities in Two Variables

Key Terms
1. boundary line 　　　2. linear inequality in two variables

Objective 1
Now Try

2.

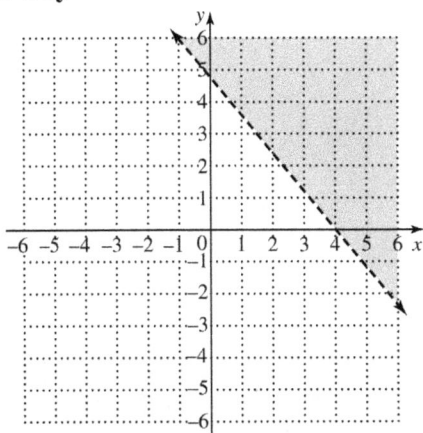

3.

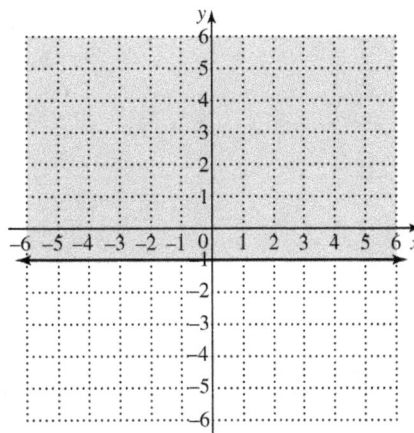

Practice Exercises

1.

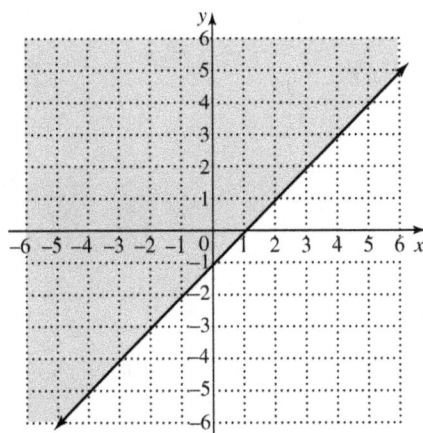

3.

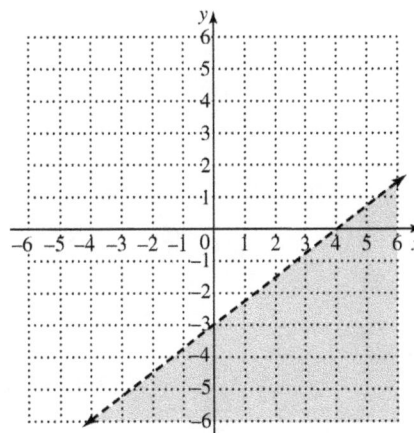

Answers

Objective 2

Now Try

4.

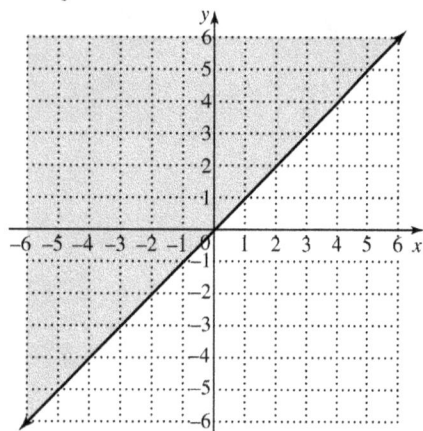

Practice Exercises

5.

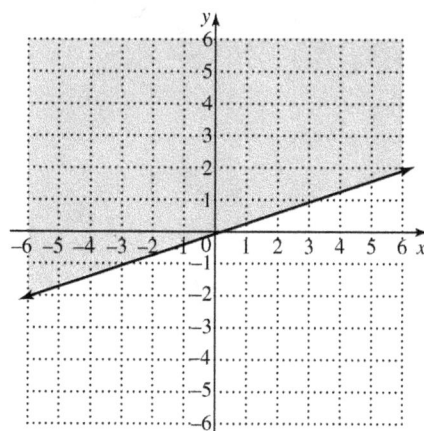

Chapter 4 SYSTEMS OF LINEAR EQUATIONS AND INEQUALITIES

4.1   Solving Systems of Linear Equations by Graphing

Key Terms
1. independent equations
2. consistent system
3. solution set of a system
4. solution of a system
5. dependent equations
6. inconsistent system
7. system of linear equations

Objective 1
Now Try
1b. no

Practice Exercises
1. no   3. no

Objective 2
Now Try   **Practice Exercises**

2.   5.

Objective 3
Now Try
3a. $\{(x, y)\mid 4x - 2y = 8\}$ 3b. $\varnothing$

Practice Exercises
7. no solution

Objective 4
Now Try
4c. neither parallel nor same line; exactly one solution

Practice Exercises
9.   (a) neither   (b) intersecting lines   (c) one solution

11.   (a) dependent   (b) one line   (c) infinitely many solutions

Answers

4.2   Solving Systems of Linear Equations by Substitution

Key Terms
  1. ordered pair   2. substitution   3. dependent system

  4. inconsistent system

Objective 1
  Now Try
  1. $\{(1, 6)\}$ 2. $\{(9, -4)\}$ 3. $\{(8, -2)\}$

  Practice Exercises
  1. $(2, 4)$ 3. $(4, -9)$

Objective 2
  Now Try
  4. $\varnothing$ 5. $\{(x, y) \mid 5x + 4y = 20\}$

  Practice Exercises
  5. $\{(x, y) \mid -x + 2y = 6\}$

Objective 3
  Now Try
  6. $\{(-2, 5)\}$

  Practice Exercises
  7. $(-9, -11)$ 9. $\{(x, y) \mid 0.6x + 0.8y = 1\}$

4.3   Solving Systems of Linear Equations by Elimination

Key Terms
  1. elimination method   2. addition property of equality

  3. substitution

Objective 1
  Now Try
  1. $\{(8, 3)\}$

  Practice Exercises
  1. $(8, 3)$ 3. $(5, 0)$

Objective 2
  Now Try
  4. $\{(-4, 9)\}$ 3. $\{(1, -3)\}$

  Practice Exercises
  5. $\left(\dfrac{1}{2}, 1\right)$

Objective 3
Now Try

5. $\left\{ \left(\dfrac{9}{17}, \dfrac{11}{17} \right) \right\}$

Practice Exercises

7. $(2, -4)$ 9. $(3, -2)$

Objective 4
Now Try

6a. $\varnothing$ 6b. $\left\{ (x, y) \mid 9x - 7y = 5 \right\}$

Practice Exercises

11. $\left\{ (x, y) \mid -x - 2y = 3 \right\}$

4.4 Applications of Linear Systems

Key Terms

1. $d = rt$ 2. system of linear equations

Objective 1
Now Try

1. 56 cm, 26 cm

Practice Exercises

1. 32, 18 3. length: 14 ft; width: 11 ft

Objective 2
Now Try

2. 1500 general admission tickets; 750 reserved seats

Practice Exercises

5. 30 $5 bills; 60 $10 bills

Objective 3
Now Try

3. 32 oz of 65% solution; 48 oz of 85% solution

Practice Exercises

7. $6 coffee: 100 lbs; $12 coffee: 50 lb

9. $1.60 candy: 20 lb; $2.50 candy: 10 lb

Objective 4
Now Try

4. Enid: 44 mph; Jerry: 16 mph

Practice Exercises

11. plane A: 400 mph; plane B: 360 mph

4.5 Graphing Linear Inequalities in Two Variables

Key Terms
1. solution set of a system of linear inequalities   2. system of linear inequalities

Objective 1
Now Try

1.

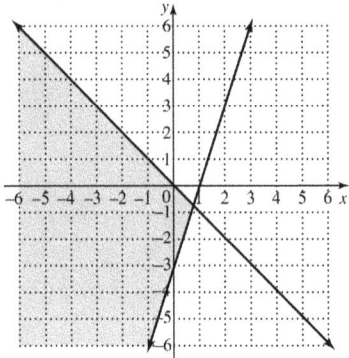

2.

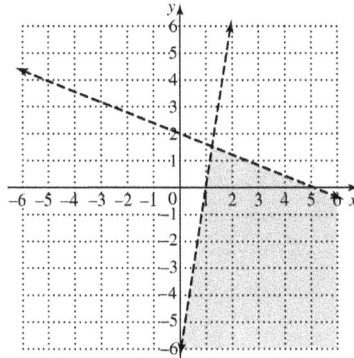

3.

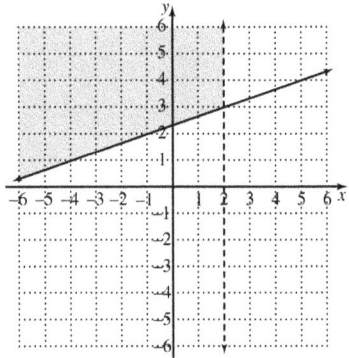

Practice Exercises

1.

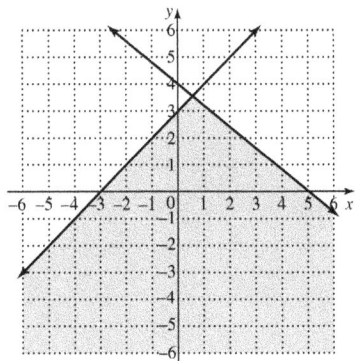

3.

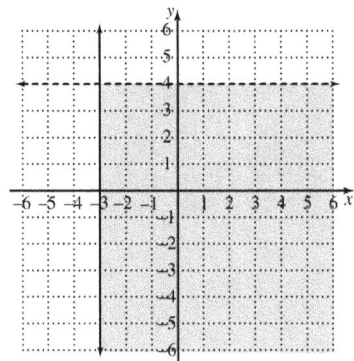

Chapter 5 EXPONENTS AND POLYNOMIALS

5.1   The Product Rule and Power Rules for Exponents

Key Terms
1. power   2. exponential expression   3. base

Objective 1
   Now Try
   1. 4^5; 1024   2a. base: 2; exponent: 6; value: 64

2b. base:2; exponent: 6; value: –64

2c. base: –2; exponent: 6; value 64

   Practice Exercises
   1. $\left(\dfrac{1}{3}\right)^5$ 3. –6561; base: 3; exponent: 8

Objective 2
   Now Try
3a. 9^{13} 3d. m^{27} 4. $18x^{11}$

3f. 108

   Practice Exercises
   5. $8c^{15}$

Objective 3
   Now Try
5a. 4^{15} 5c. x^{30}

   Practice Exercises
   7. 7^{12} 9. $(-3)^{21}$

Objective 4
   Now Try
6a. $64a^3b^3$

   Practice Exercises
11. $-0.008a^{12}b^3$

Objective 5
   Now Try
7b. $\dfrac{1}{1024}$

Answers

Practice Exercises

13. $-\dfrac{8x^3}{125}$

15. $-\dfrac{128a^7}{b^{14}}$

Objective 6
Now Try

8a. $\dfrac{5^5}{2^3}$, or $\dfrac{3125}{8}$

8d. $-x^{27}y^{13}$

Practice Exercises

17. $32a^9b^{14}c^5$

Objective 7
Now Try

9a. $28x^5$

Practice Exercises

19. $36x^5$

21. $28q^{11}$

5.2   Integer Exponents and the Quotient Rule

Key Terms

1. power rule for exponents

2. base; exponent

3. product rule for exponents

Objective 1
Now Try

1a. 1   1b. 1   1c. −1   1d. 1   1e. 88

1f. 1

Practice Exercises

1. −1   3. 0

Objective 2
Now Try

2a. $\dfrac{1}{64}$ 2d. $\dfrac{49}{36}$ 2f. $\dfrac{1}{8}$ 2h. x^8 3b. $\dfrac{y}{x^7}$

3c. $\dfrac{qr^5}{4p^3}$ 3d. $\dfrac{625b^4}{a^4}$

Practice Exercises

5. $\dfrac{1}{m^{18}n^9}$

Objective 3
 Now Try

4b. $\dfrac{1}{16}$ 4d. z^{10} 4e. $\dfrac{a^3}{25}$

 Practice Exercises

7. $\dfrac{k^4 m^5}{2}$ 9. $\dfrac{p^8}{3^5 m^3}$ or $\dfrac{p^8}{243 m^3}$

Objective 4
 Now Try

5d. $\dfrac{243}{32 p^{20}}$ 5e. $\dfrac{x^{12} y^{15}}{8000}$

 Practice Exercises

11. $a^{16} b^{22}$

5.3   An Application of Exponents: Scientific Notation

Key Terms
 1. scientific notation   2. power rule   3. quotient rule

Objective 1
 Now Try

1b. 4.771×10^{10} 1c. 4.63×10^{-2}

 Practice Exercises

 1. 2.3651×10^4 3. -2.208×10^{-4}

Objective 2
 Now Try

2b. 27,960,000   2d. -0.000164

 Practice Exercises

 5. 0.0064

Objective 3
 Now Try

3a. 2.7×10^8, or 270,000,000   3b. 3×10^{-8}, or 0.00000003

 4. 5.45×10^3 kg/m^3

 Practice Exercises

 7. 2.53×10^2 9. 4.86×10^{19} atoms

Answers

5.4   Adding and Subtracting Polynomials

Key Terms
1. degree of a term
2. descending powers
3. term
4. trinomial
5. polynomial
6. monomial
7. degree of a polynomial 8. binomial
9. like terms

Objective 1
Now Try
1a. $-2, -1, 3; \ 3$ 1b. $-8, 1; \ 2$

Practice Exercises

1. $1; \ 1$ 3. $\dfrac{2}{5}, -3, 1; \ 3$

Objective 2
Now Try
2e. $15m^3 + 29m^2$ 2b. $-8x^7$

Practice Exercises
5. $2.6z^8 - 0.9z^7$

Objective 3
Now Try
3a. $8x^3 + 4x^2 + 6$; degree 3; trinomial   3d. $4x^5$; degree 5; monomial

Practice Exercises
7. $n^8 - n^2$; degree 8; binomial
9. $5c^5 + 3c^4 - 10c^2$; degree 5; trinomial

Objective 4
Now Try
4a. 1285 4b. 1357

Practice Exercises
11. a. 71; b. -19

Objective 5
Now Try
5a. $5x^3 - 5x^2 + 4x$ 6b. $7x^3 + 8x^2 - 14x - 1$

Practice Exercises
13. $5m^3 - 2m^2 - 4m + 4$ 15. $3r^3 + 7r^2 - 5r - 2$

Objective 6
Now Try
7b. $-11x^3 - 7x + 1$

Practice Exercises

17. $8b^4 - 4b^3 - 2b^2 - b - 2$

Objective 7
Now Try

10b. $x^2y + 4xy$

Practice Exercises

19. $-3a^6 + a^4b - 3b^2$ 21. $-7x^2y + 3xy + 5xy^2$

5.5   Multiplying Polynomials

Key Terms

1. inner product   2. FOIL   3. outer product

Objective 1
Now Try

1a. $-72p^4$

Practice Exercises

1. $10m^4$ 3. $-27a^4b^3$

Objective 2
Now Try

2a. $32x^4 + 64x^3$

Practice Exercises

5. $6m + 14m^3 + 6m^4$

Objective 3
Now Try

3. $4x^7 - 2x^5 + 37x^4 - 18x^2 + 9x$
4. $28x^4 - 33x^3 + 51x^2 + 17x - 15$

Practice Exercises

7. $x^3 + 27$ 9. $6x^4 + 11x^3 - 9x^2 - 4x$

Objective 4
Now Try

5. $x^2 + 3x - 54$ 6. $16xy + 72y - 14x - 63$
7b. $45p^2 + 11pq - 4q^2$

Practice Exercises

11. $3 + 10a + 8a^2$

Answers

5.6 Special Products
Key Terms
1. binomial   2. conjugate

Objective 1
Now Try

1. $x^2 + 12x + 36$ 2b. $64a^2 - 48ab + 9b^2$ 2c. $4a^2 + 36ak + 81k^2$

Practice Exercises

1. $49 + 14x + x^2$ 3. $16y^2 - 5.6y + 0.49$

Objective 2
Now Try

3a. $x^2 - 81$ 4c. $x^2 - \dfrac{9}{25}$ 4b. $121x^2 - y^2$

4d. $4p^5 - 144p$

Practice Exercises

5. $64k^2 - 25p^2$

Objective 3
Now Try

5a. $x^3 + 18x^2 + 108x + 216$

5b. $81x^4 - 540x^3 + 1350x^2 - 1500x + 625$

Practice Exercises

7. $a^3 - 9a^2 + 27a - 27$
9. $256s^4 + 768s^3t + 864s^2t^2 + 432st^3 + 81t^4$

5.7 Dividing a Polynomial by a Monomial
Key Terms
1. dividend   2. quotient   3. divisor

Objective 1
Now Try

1. $x^2 - 3x$ 2. $3n^2 - 4n - \dfrac{2}{n}$ 3. $\dfrac{7z^4}{2} - 4z^3 - \dfrac{5}{z} - \dfrac{3}{z^2}$

4. $-2a^3b^2 - 4a^2b + 3$

Practice Exercises

1. $2a^3 - 3a$ 3. $-13m^2 + 4m - \dfrac{5}{m^2}$

5.8   Dividing a Polynomial by a Polynomial

Key Terms

1. divisor   2. quotient   3. dividend

Objective 1
Now Try

2. $2x^2 - 2x - 2 + \dfrac{-5}{5x - 1}$

3. $x^2 + 10x + 100$

4. $3x^2 + 5x + 5 + \dfrac{8x + 29}{x^2 - 4}$

5. $x^2 + \dfrac{1}{8}x + \dfrac{5}{8} + \dfrac{6}{8x - 8}$

Practice Exercises

1. $-3x + 4$ 3. $a^2 + 1$

Objective 2
Now Try

6. $L = 2r^2 - r + 5$ units

Practice Exercises

5. $3t + 4$ units

Chapter 6 FACTORING AND APPLICATIONS

6.1   Greatest Common Factors; Factor by Grouping

Key Terms

  1. factoring   2. factored form   3. greatest common factor

  4. factor

Objective 1
  Now Try

1b. 8   1c. 1

  Practice Exercises

1. 28   3. 6

Objective 2
  Now Try

2a. $6x^4$ 2b. q^3

  Practice Exercises

5. $k^2 m^4 n^4$

Objective 3
  Now Try

3a. $8x\left(x^4 + 3\right)$ 3b. $4y^2\left(5y^2 - 3y + 1\right)$ 5a. $(y + 8)(y + 4)$

  Practice Exercises

7. $10x\left(2x + 4xy - 7y^2\right)$ 9. $-13x^8\left(-2 + x^4 - 4x^2\right)$

Objective 4
  Now Try

6a. $(2 + y)(x + 7)$ 6c. $(x - 7)(4x + 5y)$

  Practice Exercises

11. $(2x + y)(x - 7y)$

6.2   Factoring Trinomials

Key Terms

  1. factoring   2. greatest common factor   3. prime polynomial

Objective 1
  Now Try

2. $(y - 7)(y - 5)$ 4. $(a - 17)(a + 2)$ 5a. prime

6. $(p - 7q)(p + 2q)$

Practice Exercises
1. prime   3. $(x-11)(x+3)$

Objective 2
  Now Try
  7. $7x^4(x-5)(x-2)$

Practice Exercises
  5. $2ab(a-3b)(a-2b)$

6.3   Factoring Trinomials by Grouping

Key Terms
  1. coefficient   2. trinomial

Objective 1
  Now Try
  2a. $(7x-5)(2x+1)$ 2b. $(3m-7)(m+2)$ 2c. $(5x+3y)(2x-y)$

  3. $3x^3(5x-3)(2x+7)$

Practice Exercises
  1. $(4b+3)(2b+3)$ 3. $(5c-7t)(2c-3t)$

6.4   Factoring Trinomials Using the FOIL Method

Key Terms
  1. inner product   2. FOIL   3. outer product

Objective 1
  Now Try
  1. $(3x+2)(x+5)$ 2. $(3x+1)(5x+7)$ 3. $(4x-1)(5x-2)$

  4. $(4x+7)(2x-3)$ 5. $(6x-5y)(4x+3y)$

Practice Exercises
  1. $(2q+1)(4q+3)$ 3. $2(2c-d)(c+4d)$

6.5 Special Factoring Techniques

Key Terms
  1. difference   2. perfect square trinomial

Objective 1
  Now Try
  1b. prime   2a. $(2x+9)(2x-9)$ 3a. $10(3x+7)(3x-7)$
  3b. $(x^2+8)(x^2-8)$ 3c. $(p^2+16)(p+4)(p-4)$

  Practice Exercises
  1. $(x-7)(x+7)$ 3. prime

Objective 2
  Now Try
  4. $(p+8)^2$ 5d. $5x(2x+5)^2$

  Practice Exercises
  5. $(3j+2)^2$

6.6 Solving Quadratic Equations using the Zero-Factor Property

Key Terms
  1. standard form   2. quadratic equation

Objective 1
  Now Try
  1a. $\left\{-12, \frac{7}{4}\right\}$ 2a. $\{-3, 8\}$ 3. $\left\{\frac{4}{5}, 3\right\}$

  4c. $\left\{-\frac{1}{4}, 6\right\}$ 4b. $\{0, 11\}$

  Practice Exercises
  1. $\left\{-\frac{5}{2}, 4\right\}$ 3. $\left\{-4, \frac{3}{5}\right\}$

Objective 2
  Now Try
  6. $\{-5, 0, 5\}$

  Practice Exercises
  5. $\{-9, 0, 1\}$

6.7   Applications of Quadratic Equations

Key Terms
1. legs   2. hypotenuse

Objective 1
Now Try
1. width: 3 m, length: 5 m

Practice Exercises
1. width: 8 in., length: 24 in.   3. height: 4 ft, width: 6 ft

Objective 2
Now Try
3. 8, 10   2. −4, −3 or 3, 4

Practice Exercises
5. 6, 8

Objective 3
Now Try
4. 16 ft

Practice Exercises
7. 45 m, 60 m, 75 m   9. 20 mi

Objective 4
Now Try
5. 1 sec

Practice Exercises
11. 40 items or 110 items

Chapter 7 FACTORING AND APPLICATIONS

7.1　The Fundamental Property of Rational Expressions

Key Terms
　1. rational expression　　2. lowest terms

Objective 1
　Now Try

1a. $\dfrac{3}{2}$　　　　　　1b. $\dfrac{7}{5}$　　　　　　1c. undefined

1d. 0

　Practice Exercises

1. a. $-\dfrac{11}{9}$; b. -4　　3. a. $-\dfrac{1}{6}$; b. $-\dfrac{7}{2}$

Objective 2
　Now Try

2a. $y \neq \dfrac{1}{7}$　　　　　2b. $m \neq 5,\ m \neq -4$　　　　2c. never undefined

　Practice Exercises

5. none

Objective 3
　Now Try

3b. $\dfrac{3}{k^3}$　　　　　　4a. $\dfrac{7}{9}$　　　　　　4c. $\dfrac{m+6}{2m+3}$

5. -1

　Practice Exercises

7. $\dfrac{-5b}{8c}$　　　　　　9. $\dfrac{9(x+3)}{2}$

Objective 4
　Now Try

7. $\dfrac{-(10x-7)}{4x-3}$, $\dfrac{-10x+7}{4x-3}$, $\dfrac{10x-7}{-(4x-3)}$, $\dfrac{10x-7}{-4x+3}$

　Practice Exercises

11. $\dfrac{-(2p-1)}{-(1-4p)}$; $\dfrac{-2p+1}{-1+4p}$; $\dfrac{-(2p-1)}{-1+4p}$; $\dfrac{-2p+1}{-(1-4p)}$

7.2   Multiplying and Dividing Rational Expressions

Key Terms
1. reciprocal   2. rational expression   3. lowest terms

Objective 1
Now Try

1b. $\dfrac{4}{3x}$ 2. $\dfrac{s^2}{6(r-s)}$

Practice Exercises

1. $\dfrac{10m^3n}{3}$ 3. $\dfrac{x+4}{2(x-4)}$

Objective 2
Now Try

4a. $\dfrac{10b}{11a^3}$ 4b. $\dfrac{w^2+6w-7}{w^2-6w}$

Practice Exercises

5. $\dfrac{1}{4r^2+2r+3}$

Objective 3
Now Try

6. $\dfrac{2}{25}$ 7. $\dfrac{8x}{(x-3)^2}$ 8. $\dfrac{-(m-8)}{5m(m+9)}$

Practice Exercises

7. $-\dfrac{1}{2}$ 9. $\dfrac{-3(6k-1)}{3k-1}$

7.3   Least Common Denominators

Key Terms
1. equivalent expressions   2. least common denominator

Objective 1
Now Try

2. $120a^4$ 3a. $9w(w-2)$ 3b. $(b+4)(b+1)(b-4)^2$

Practice Exercises

1. $108b^4$ 3. $w(w+3)(w-3)(w-2)$

Answers

Objective 2
  Now Try
  4a. $\dfrac{65}{30}$ 5a. $\dfrac{76}{24c-20}$

  5b. $\dfrac{3(z+2)}{z(z-7)(z+2)}$ or $\dfrac{3z+6}{z^3-5z-14z}$

  Practice Exercises
  5. $30r$

7.4   Adding and Subtracting Rational Expressions

Key Terms
  1. greatest common factor   2. least common multiple

Objective 1
  Now Try
  1b. $2x$

  Practice Exercises
  1. $\dfrac{4}{w^2}$ 3. $\dfrac{1}{x-2}$

Objective 2
  Now Try
  2a. $\dfrac{137}{315}$ 3. $\dfrac{4}{x-10}$ 4. $\dfrac{7x^2+11x+8}{(x+2)(x+1)(x-4)}$

  Practice Exercises
  5. $\dfrac{7z^2-z-6}{(z+2)(z-2)^2}$

Objective 3
  Now Try
  8. 7   9. $\dfrac{8x^2+37x+15}{(x+5)(x-5)^2}$ 10. $\dfrac{p+25}{(p+1)(p-1)(p-5)}$

  Practice Exercises
  7. $\dfrac{8z}{(z-2)(z+2)}$ or $\dfrac{8z}{z^2-4}$ 9. $\dfrac{2m^2-m+2}{(m-2)(m+2)^2}$

7.5 Complex Fractions

Key Terms

1. complex fraction 2. LCD

Objective 1

Objective 2
  Now Try

1b. $\dfrac{90}{x}$ 2. $\dfrac{bc^2}{a^2}$ 3. $\dfrac{-9x+65}{2x-5}$

  Practice Exercises

1. $\dfrac{7m^2}{2n^3}$ 3. $\dfrac{9s+12}{6s^2+2s}$ or $\dfrac{3(3s+4)}{2s(3s+1)}$

Objective 3
  Now Try

4b. $\dfrac{12}{x}$ 5. $\dfrac{5n-9}{3(6n+5)}$

  Practice Exercises

5. $\dfrac{(x-2)^2}{x(x+2)}$

7.6 Solving Equations with Rational Expressions

Key Terms

1. extraneous solution 2. proposed solution

Objective 1
  Now Try

1a. expression; $\dfrac{1}{2}x$ 1b. equation; $\{14\}$

  Practice Exercises

1. equation; $\{-2\}$ 3. expression; $\dfrac{41x}{15}$

Objective 2
  Now Try
7. $\{-4\}$ 6. $\{0\}$ 5. $\{6\}$
  Practice Exercises
5. $\{-4, 16\}$

Objective 3
  Now Try
9. $b = aq + c$
  Practice Exercises

7. $f = \dfrac{d_0 d_1}{d_0 + d_1}$ 9. $q = \dfrac{2pf - Ab}{Ab}$ or $\dfrac{2pf}{Ab} - 1$

7.7   Applications of Rational Expressions

Key Terms

1. numerator

2. denominator

3. reciprocal

Objective 1

Now Try

1. 3

Practice Exercises

1. $-\frac{2}{3}$ or 1

3. $\frac{3}{5}$

Objective 2

Now Try

4. 3 miles per hour

Practice Exercises

5. 24 miles per hour

Objective 3

Now Try

5. $\frac{2}{5}$ hour

Practice Exercises

7. $2\frac{2}{5}$ hr

9. 72 hr

7.8   Variation

Key Terms

1. constant of variation

2. direct variation

3. inverse variation

Objective 1

Now Try

1. 168

2. 69.08 cm

Practice Exercises

1. 12

3. $162.50

Objective 2

Now Try

3. 5

4. 7.2 pounds per square foot

Practice Exercises

5. 2.25

Chapter 8 ROOTS AND RADICALS

8.1   Evaluating Roots

Key Terms

1. radicand
2. perfect square
3. index (order)
4. square root
5. radical expression
6. principal square root
7. perfect cube
8. irrational number
9. radical
10. cube root

Objective 1
Now Try
1. 9, −9   2a. 13   2d. 0.8

3a. 19   3c. $n^2 + 5$

Practice Exercises
1. 25, −25   3. $\dfrac{30}{7}$

Objective 2
Now Try
4a. irrational   4b. rational   4c. not a real number

Practice Exercises
5. not a real number

Objective 3
Now Try
5c. −26.038

Practice Exercises
7. 5.657   9. 14.491

Objective 4
Now Try
6a. $c = 17$ 7. 35 miles

Practice Exercises
11. 9

Objective 5
Now Try
8a. 5   8b. −5   9a. 5

9e. −5

Practice Exercises
13. −7   15. −1

Answers

8.2 Multiplying, Dividing, and Simplifying Radicals

Key Terms

1. radical   2. perfect cube   3. radicand

Objective 1

Now Try

1b. $\sqrt{42}$ 1c. $\sqrt{15c}$

Practice Exercises

1. $\sqrt{65}$ 3. $\sqrt{21x}$

Objective 2

Now Try

2c. $4\sqrt{5}$ 3c. $60\sqrt{2}$

Practice Exercises

5. $11\sqrt{3}$

Objective 3

Now Try

4a. $\dfrac{13}{3}$ 4b. 9   4c. $\dfrac{\sqrt{7}}{8}$

5. $3\sqrt{10}$ 6. $\dfrac{\sqrt{7}}{6}$

Practice Exercises

7. $\dfrac{5}{9}$ 9. $\dfrac{2}{25}$

Objective 4

Now Try

7c. $6x^2\sqrt{2}$

Practice Exercises

11. $4x^2 y^2 \sqrt{2y}$

Objective 5

Now Try

8c. $\dfrac{2}{7}$ 9a. m^3 9c. $2a\sqrt[4]{5a}$

Practice Exercises

13. $-2\sqrt[5]{2}$ 15. $\dfrac{5}{4}$

8.3   Adding and Subtracting Radicals

Key Terms
1. index   2. unlike radicals   3. like radicals

Objective 1
   Now Try
1b. $-6\sqrt{17}$ 1c. $6\sqrt{14}$

   Practice Exercises
   1. $5\sqrt{2} + \sqrt{3}$ 3. $-5\sqrt{5}$

Objective 2
   Now Try
2b. $\sqrt{2}$ 2c. $27\sqrt{6}$

   Practice Exercises
   5. $27\sqrt{2}$

Objective 3
   Now Try
3b. $5\sqrt{7k}$ 3c. $32x\sqrt{3}$ 3d. $11m\sqrt[3]{2}$

   Practice Exercises
   7. $4\sqrt{35}$ 9. $39w\sqrt{6}$

8.4   Rationalizing the Denominator

Key Terms
1. product rule   2. rationalizing the denominator

3. quotient rule

Objective 1
   Now Try
1a. $2\sqrt{7}$ 1b. $\dfrac{\sqrt{3}}{3}$

   Practice Exercises
   1. $\dfrac{3\sqrt{10}}{2}$ 3. $\dfrac{3}{5}$

Objective 2
   Now Try
2. $\dfrac{\sqrt{21}}{6}$ 3. $\dfrac{\sqrt{10}}{15}$

Practice Exercises

5. $\dfrac{m^2\sqrt{k}}{k^2}$

Objective 3

Now Try

5b. $\dfrac{\sqrt[3]{33}}{3}$

Practice Exercises

7. $\dfrac{\sqrt[3]{18}}{3}$

9. $\dfrac{\sqrt[3]{35x^2}}{7x}$

8.5   More Simplifying and Operations with Radicals

Key Terms

1. rationalize the denominator

2. conjugate

Objective 1

Now Try

1a. $\sqrt{10}$

1b. $-69-2\sqrt{30}$

3a. 117

2b. $114+80\sqrt{2}$

Practice Exercises

1. $4\sqrt{14}-63$

3. -38

Objective 2

Now Try

4a. $\dfrac{10\left(7-\sqrt{10}\right)}{39}$

4b. $\dfrac{-8\sqrt{6}-21}{3}$

Practice Exercises

5. $\dfrac{-\sqrt{3}-2\sqrt{6}-\sqrt{2}-4}{7}$

Objective 3

Now Try

5. $\dfrac{\sqrt{5}+5}{9}$

Practice Exercises

7. $\dfrac{1+\sqrt{3}}{3}$

9. $27\sqrt{3}+5$

8.6   Solving Equations with Radicals

Key Terms
  1. extraneous solution   2. radical equation

Objective 1
  Now Try
  1. $\{11\}$ 2. $\{3\}$

  Practice Exercises
  1. $\left\{\dfrac{8}{3}\right\}$ 3. $\{2\}$

Objective 2
  Now Try
  3. $\varnothing$

  Practice Exercises
  5. $\varnothing$

Objective 3
  Now Try
  6. $\{-2, -1\}$ 5. $\{7\}$ 7. $\{5\}$

  Practice Exercises
  7. $\{8\}$ 9. $\{3\}$

Objective 4
  Now Try
  8. 6804 m^2

  Practice Exercises
11. 499 sq cm

Chapter 9 QUADRATIC EQUATIONS

9.1   Solving Quadratic Equations by the Square Root Property

Key Terms
  1. quadratic equation   2. zero-factor property

Objective 1
  Now Try
  1a. $\{-7, -1\}$

  Practice Exercises
  1. $\{-4, -2\}$ 3. $\{-7, 5\}$

Objective 2
  Now Try
  2a. $\{-9, 9\}$ 2b. $\left\{-\sqrt{23},\ \sqrt{23}\right\}$ 3. $\varnothing$
  2c. $\left\{-5\sqrt{3},\ 5\sqrt{3}\right\}$, or $\left\{\pm 5\sqrt{3}\right\}$

  Practice Exercises
  5. $\left\{-7\sqrt{2},\ 7\sqrt{2}\right\}$

Objective 3
  Now Try
  4b. $\left\{-3 \pm \sqrt{13}\right\}$ 5. $\left\{\dfrac{3 \pm 4\sqrt{2}}{7}\right\}$ 6. $\varnothing$

  Practice Exercises
  7. $\{-6, 2\}$ 9. $\left\{\dfrac{1}{5},\ \dfrac{4}{5}\right\}$

Objective 4
  Now Try
  7. About 16.9 in.

  Practice Exercises
11. 2 in.

9.2 Solving Quadratic Equations by Completing the Square

Key Terms

1. perfect square trinomial

2. square root property

3. completing the square

Objective 1

Now Try

1a. $25; (x+5)^2$ 1b. $121; (x-11)^2$ 2. $-5 \pm 4\sqrt{2}$

3. $\{-8+\sqrt{73}, -8-\sqrt{73}\}$

Practice Exercises

1. $\{-4-2\sqrt{3}, \ -4+2\sqrt{3}\}$ 3. $\{-9, 7\}$

Objective 2

Now Try

4. $\left\{\dfrac{5}{4}, \dfrac{11}{4}\right\}$ 6. $\varnothing$

Practice Exercises

5. $\varnothing$

Objective 3

Now Try

7. $\{-3+\sqrt{30}, -3-\sqrt{30}\}$

Practice Exercises

7. $\varnothing$ 9. $\{-2-\sqrt{2}, \ -2+\sqrt{2}\}$

Objective 4

Now Try

9. 3 sec

Practice Exercises

11. 2 months, 4 months

Answers

9.3   Solving Quadratic Equations by the Quadratic Formula

Key Terms
  1. standard form   2. constant   3. quadratic formula

Objective 1
  Now Try
  1a. $a = 12, b = -10, c = 7$ 1b. $a = -4, b = -110, c = 8$
  1e. $a = 30, b = 11, c = -39$

  Practice Exercises
  1. $a = 10, b = 4, c = 0$ 3. $a = 1, b = 3, c = 9$

Objective 2
  Now Try
  2. $\left\{-2, \dfrac{7}{3}\right\}$ 3. $\left\{3 + \sqrt{5}, \ 3 - \sqrt{5}\right\}$

  Practice Exercises
  5. $\varnothing$

Objective 3
  Now Try
  5. $\left\{\dfrac{11}{2}\right\}$

  Practice Exercises
  7. $\{2\}$ 9. $\left\{\dfrac{9}{7}\right\}$

Objective 4
  Now Try
  6. $\varnothing$

  Practice Exercises
  11. $\varnothing$

9.4 Graphing Quadratic Equations

Key Terms

1. line of symmetry   2. vertex   3. axis

4. parabola

Objective 1

Now Try

2.

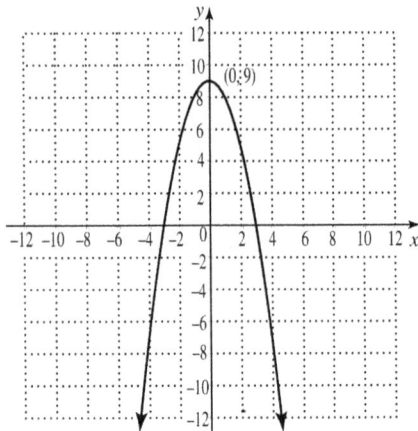

Objective 2

Now Try

3.

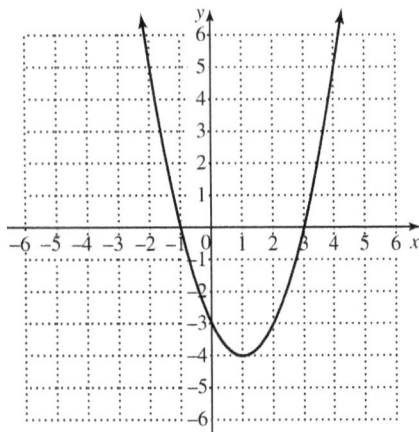

4.

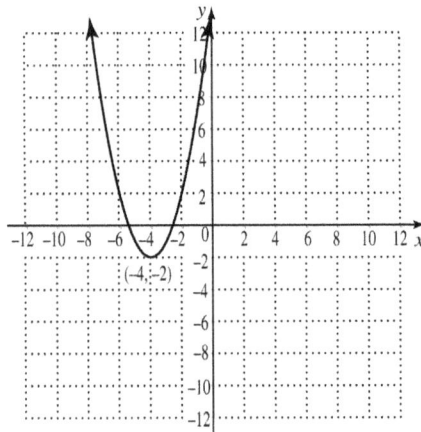

435

Answers

Objective 1 & 2

Practice Exercises

1.

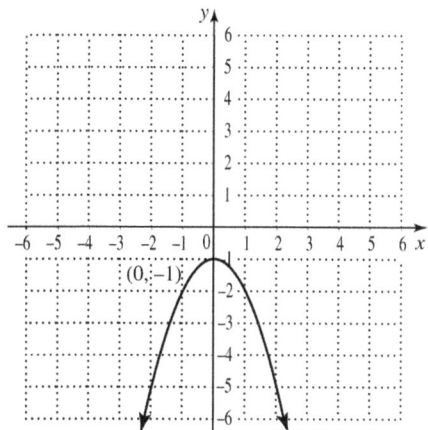

Vertex: $(0, -1)$

3.

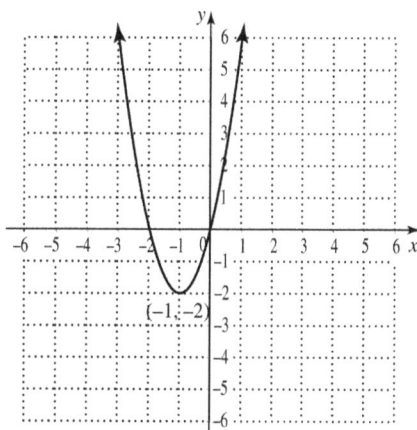

Vertex: $(-1, -2)$

5.

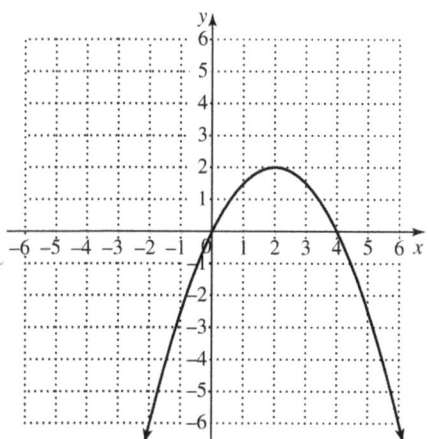

Vertex: $(2, 2)$

9.5   Introduction to Functions

Key Terms
  1. relation   2. range   3. function
  4. components   5. domain

Objective 1
  Now Try
  1a. domain: {6, 8, 10, 12}; range: {7, 9, 11, 13}
  1b. domain: {1}; range: {11, 12, 13, 14}

  Practice Exercises
  1. domain: {−3, 0, 2, 5}; range: {−8, −4, −1, 2, 7}
  3. domain: {−3, −2, −1, 0, 1}; range: {−5, 0, 5}

Objective 2
  Now Try
  2a. function   2b. not a function

  Practice Exercises
  5. not a function

Objective 3
  Now Try
  3c. not a function   3d. function

  Practice Exercises
  7. not a function   9. function

Objective 4
  Now Try
  4a. 5   4b. −10   4c. −15

  Practice Exercises
  11. (a) 6; (b) 2; (c) 18

Objective 5
  Now Try
  5a. {(2003, 719 million), (2004, 817 million), (2005, 1018 million),
       (2006, 1093 million), (2007, 1262 million)}; yes
  5b. Domain: {2003, 2004, 2005, 2006, 2007}
       Range: {719 million, 817 million, 1018 million, 1093 million, 1262 million }
  5c. $w(2003) = 719$ million; $w(2006) = 1093$ million   5d. 2005

  Practice Exercises
  13. {(2008, 596 thousand), (2009, 625 thousand), (2010, 872 thousand),
       (2011, 795 thousand), (2012, 625 thousand)}; function
  15. $r(2009) = \$625$ thousand; $r(2011) = \$795$ thousand

www.ingramcontent.com/pod-product-compliance
Lightning Source LLC
Chambersburg PA
CBHW081730220326

R18016800001B/R180168PG41598CBX00001B/1